MANUAL OF
SOLID STATE
CIRCUIT DESIGN AND
TROUBLESHOOTING

# MANUAL OF SOLID STATE CIRCUIT DESIGN AND TROUBLESHOOTING

VESTER ROBINSON

Reston Publishing Company, Inc., Reston, Virginia
*A Prentice-Hall Company*

Library of Congress Cataloging in Publication Data

Robinson, Vester.
  Manual of solid state circuit design and troubleshooting.

  Bibliography:  p.
  Includes index.
  1. Electronic circuit design.   2. Semiconductors.
3. Electronic circuits—Testing.   4. Semiconductors—
Testing.   I. Title.
TK7867.R6            621.3815'3'042          76-26555
ISBN 0-87909-464-8

© 1977 by Reston Publishing Company, Inc.
*A Prentice-Hall Company*
11480 Sunset Hills Road
Reston, Virginia 22090


All rights reserved. No part of this book
may be reproduced in any way, or by any means,
without permission in writing from the publisher.


10  9  8  7  6  5  4  3  2  1

Printed in the United States of America

*TO FLORENCE BELLE*

# Contents

Preface xi

### Section One
## CHARACTERISTICS AND BIASING

#### Chapter One
### Diodes and Thyristors   3

P-N Junction Diode, 3   Zener Diode, 8   Tunnel Diode, 10   Tunnel Rectifier, 12   Varactor Diode, 14   Silicon-Controlled Rectifier, 15   Triac, 17   Diode and Thyristor Symbols, 21

#### Chapter Two
### Bipolar Transistors   24

Structure, 24   Controlling the Electron Flow, 26   Basic Circuits, 29   Practical Biasing Arrangements, 36   Transistor Tests, 38   Common Symbols, 44

#### Chapter Three
### Field-Effect Transistors   49

Junction Field-Effect Transistor, 49   Insulated Gate Field-Effect Transistor, 58   Basic FET Circuits, 65   JFET Biasing, 72   MOSFET Biasing, 74   FET Symbols, 77

## Section Two
# DESIGN CONSIDERATIONS

### Chapter Four
### Interpreting Data Sheets  81

Sample Data Sheet, 81   Actual Sample Data Sheet, 86

### Chapter Five
### Common Problems in Design  99

Basic Parameters, 99   Parameter Conversion, 105   Application of Parameters, 109   Effects of Frequency on Parameters, 115   Effects of Temperature on Parameters, 117   Use of Heat Sinks, 118   Thermal Circuits, 118   Test Equipment, 122   Spare Parts, 124

## Section Three
# AUDIO-FREQUENCY CIRCUITS

### Chapter Six
### Bipolar Voltage Amplifiers  127

Basic Bipolar Amplifier, 127   Hybrid Equivalent Analysis, 132   Deriving Logical Equations, 135   Load-Line Analysis, 140   Design Example, 149   Testing and Troubleshooting, 152

### Chapter Seven
### Unipolar Voltage Amplifiers  155

Basic JFET Amplifier, 155   JFET Equivalent-Circuit Analysis, 158   JFET Load-Line Analysis, 162   Basic MOSFET Amplifier, 165   MOSFET Equivalent-Circuit Analysis, 168   MOSFET Load-Line Analysis, 170   Test Circuits, 172   Design Example, 172   Testing and Troubleshooting, 174

### Chapter Eight
### Power Amplifiers  176

Power Limitations, 176   Class A Amplifier with a Resistive Load, 180   Class A Amplifier with a Transformer Load, 182   Distortion in a Class A Power Amplifier, 184   Class B Power Amplifier, 186   Class AB Power Amplifier, 189   Design Example, 190   Troubleshooting, 194

Contents

### Chapter Nine
### Noise in Audio Amplifiers  197

Signal-to-Noise Ratio, 196   Noise Spectrum, 199   Noise Factor, 200   Equivalent Noise Generators, 203   Bipolar Transistor Noise, 204   Field-Effect-Transistor Noise, 205

### Chapter Ten
### Multistage Amplifiers  208

Resistor-Capacitor Coupling, 208   Transformer Coupling, 214   Direct Coupling, 220   Using Discrete Components, 220   Hybrid Packages, 222

### Chapter Eleven
### Audio Feedback  227

Feedback Principles, 227   Control Circuits, 233   Results of Feedback, 240


## Section Four
## RADIO-FREQUENCY CIRCUITS

### Chapter Twelve
### Radio-Frequency Amplifiers  245

Resonant Circuits, 245   Radio-Frequency Power Amplifiers, 250   Radio-Frequency Preamplifiers, 254   IF Amplifiers, 257   Video Amplifiers, 259

### Chapter Thirteen
### Modulation and Demodulation  265

Amplitude Modulators, 265   Frequency Modulators, 268   Frequency Multipliers, 271   Mixers, 272   Frequency Converters, 276   Demodulators, 278


## Section Five
## SIGNAL-GENERATING CIRCUITS

### Chapter Fourteen
### Sinusoidal Oscillators  287

Principles of Oscillation, 287   Keyed Audio Oscillator, 289   Twin-T Audio Oscillator, 291   Lag-Line Oscillator, 293   Bipolar Colpitts RF Oscillator, 295   Crystal-Controlled Colpitts RF Oscillator, 298   MOSFET Colpitts RF Oscillator, 300   Armstrong RF Oscillator, 303   Crystal-Controlled Armstrong RF Oscillator, 305

### Chapter Fifteen
### Relaxation Oscillators   307

RC Transients, 307   Unijunction Sawtooth Generator, 309   Zener Diode Sawtooth Generator, 311   Blocking Oscillator, 312   Astable Multivibrator, 315   Monostable Multivibrator, 317   Bistable Multivibrator, 319

## Section Six
# POWER SUPPLIES

### Chapter Sixteen
### Conversion Circuits   327

Power Transformer, 327   Power Rectifiers, 329   Half-wave Rectifier, 330   Full-wave Rectifier, 332   Bridge Rectifier, 333   Power Supply Filters, 334   Voltage Multipliers, 335   Voltage Dividers, 338

### Chapter Seventeen
### Voltage Regulators   343

Zener Diode Regulator, 343   Electronic Shunt Regulator, 345   Basic Series Regulator, 347   Electronic Series Regulator, 350   Feedback Regulator, 352   Dual-Output Regulator, 355   Voltage Regulator for Heavy Current, 356

## Section Seven
# TESTING AND TROUBLESHOOTING

### Chapter Eighteen
### Bench Testing   361

Measuring Circuit Characteristics, 361   Measuring Switching Characteristics, 367   Audio Circuits, 369   Measuring Magnetic Components, 373   Measuring Antennas, 376

### Chapter Nineteen
### Troubleshooting Techniques   380

Six-Step Procedure, 380   Troubleshooting Charts, 393   Malfunction Analysis, 393

### Bibliography   405

### Index   407

# Preface

This book is intended as a guide for designing, constructing, testing, and troubleshooting practical solid state electronic circuits. It has been written to assist anyone who works with or experiments with solid state circuits. It will be of use to engineers, technicians, teachers, students, radio amateurs, hobbyists, and others interested in this type of circuit.

Section One, Characteristics and Biasing, contains rather basic information. Many of you may want to bypass this section. However, if you are not thoroughly familiar with the characteristics of all diodes and transistors you will find this section to be very helpful. It is included for the benefit of those who do not have this knowledge and for others who might need an occasional reference.

It is assumed that the user of this book has a working knowledge of high school algebra and understands the basic concepts of current, voltage, and impedance. Total avoidance of mathematics was impossible, but it has been held to a low level.

Circuit design is a science based on facts and trial and error. We must first consider all the known facts, design our circuit on paper, then modify as necessary for each particular situation. Differences in tolerances and characteristics of circuit components make each circuit new and different from all previous circuits. Even the design examples used in this book will react slightly differently for each individual who builds the circuits, and, for that matter, for the same individual who builds the circuits more than once. An attempt has been made to provide enough information to enable each person to analyze his or her particular problems and modify and correct the circuits accordingly. The samples are limited to a few common types of circuits, but it is hoped that the book will impart enough general information to allow you to expand upon these samples and design your own circuits.

I extend my appreciation to many people for their assistance in compiling this book: to the experts in many electronic firms who have been extremely generous about supplying technical data, especially the Semiconductor Products Department of General Electric; to Matthew I. Fox, the president of Reston Publishing Company, and to his editors and artists who have been regularly helpful with their professional guidance and assistance; and to Matt Mandl and David Summers who reviewed the first draft of this manuscript. These two, in particular, used their considerable technical knowledge and careful analysis to make many valuable suggestions to the author, thus greatly improving the book.

VESTER ROBINSON

MANUAL OF
# SOLID STATE CIRCUIT DESIGN AND TROUBLESHOOTING

SECTION ONE

# CHARACTERISTICS AND BIASING

The materials that compose solid-state devices are called *semiconductors*. As the name implies, these materials are neither good conductors nor good insulators. Rather they are grouped, according to their ability to conduct current, in the central zone between conductors and insulators.

The semiconductor materials found most frequently in diodes and transistors are silicon and germanium. Both silicon and germanium are very poor conductors while they are in their pure state. However, other materials can be added to either of these while they are in the molten state to alter their characteristics.

By adding certain types of impurities, crystals can be grown that have either an excess of electrons or a shortage of electrons. Material with an excess of electrons is called *N-type* or *donor material*. Material that has a deficiency of electrons is known as *P-type* or *acceptor material*.

CHAPTER ONE

# Diodes and Thyristors

When a piece of $N$-type material is chemically joined to a piece of $P$-type material, the junction takes on rectifying properties, and a *diode* is formed. There are many types of diodes, and they are named according to specific design and intended application. There are silicon rectifier diodes, Zener diodes, signal diodes, noise diodes, and many others. A diode is a two-lead single-junction device.

By contrast, a *thyristor* is a three-lead device that has two or more $PN$ junctions. The term "thyristor" is used because these devices are switched on and off in a manner similar to the action of a thyratron tube. There are several types of thyristors, but all are electronic switches. Some of these switches are activated by voltage and some by light.

## 1.1 *PN* Junction Diode

The symbol for a junction diode is shown in Figure 1-1. Notice that the section with the arrow is the *anode*. The anode is made of $P$-type material. The section with the bar joining the arrow is the *cathode*. The cathode is made of $N$-type material. All symbols for diodes, transistors, and thyristors contain an arrow. In every case, the arrow points the direction of conventional current. Electron current is, of course, in the opposite direction. Electrons flow against the arrow.

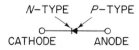

Figure 1-1. Junction-Diode Symbol

### 1.1.1 Forward Bias

A diode will conduct when the anode is made positive with respect to the cathode. This condition is called *forward bias,* and it is illustrated in Figure 1-2. Incidentally, the resistor $R1$ is an important part of this

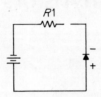

Figure 1-2.   Forward-Biased Diode

circuit. A forward-biased diode has very little resistance, and even a small quantity of voltage can cause excessive current. The resistor limits the current to a safe value and protects the diode. In this circuit, the electrons leave the negative terminal of the battery and flow through the resistor to the cathode of the diode. Electrons flow through the diode from cathode to anode with practically no opposition and return to the positive terminal of the battery.

### 1.1.2 Reverse Bias

A diode offers great opposition to current in the reverse direction. When the cathode is more positive than the anode, no current flows except a minute leakage current. This polarity of voltage constitutes *reverse bias,* and it is illustrated in Figure 1-3. With reverse bias, a few

Figure 1-3.   Reverse-Biased Diode

electrons pass through the diode from anode to cathode. This is leakage current. In most cases, this leakage is so small that it is completely ignored.

### 1.1.3 Current–Voltage Characteristics

The current through a forward-biased diode is largely controlled by the circuit. The diode in this condition is almost a direct short. That is why it is imperative to have a current-limiting device in the circuit. The relationship of current through and voltage across a diode is illustrated in Figure 1-4. When forward-bias voltage is first applied, current starts and increases very slowly to point A on the graph. When voltage increases beyond this point, a very slight change in voltage causes a very large change in current. The quantity of current for a given level of voltage is easily determined. Suppose that voltage $V$ is applied across the diode represented in Figure 1-4. In order to find the current, it is only necessary to draw a vertical line from $V$ until it intersects the curve; then draw a horizontal line from that point to the current line.

Design problems may be simplified by using an approximate characteristic curve. Since there is a negligible current below point A on the curve, it may be considered that no current flows until the voltage reaches this point. In other words, this is the *cut-in voltage*. After point A, the curve is fairly linear, and the vertical rise is very steep. The curve can be approximated as illustrated in Figure 1-5. Taking information from the approximate curve, it is considered that voltage level A is required in order to raise the current above the zero level. After the current starts, it rises nearly perpendicularly. With a silicon rectifier, voltage level A will range from 0.2 volt (V) to 0.4 V. With germanium rectifiers, this level ranges from 0.6 V to 0.7 V. These are approximations, but they are accurate enough for most common circuits. In fact, at 25 degrees Celsius (°C) (room temperature) these figures are very accurate. Above 25°C, point A moves toward the left at the rate of 2.5 millivolts (mV) for each 1°C rise in temperature.

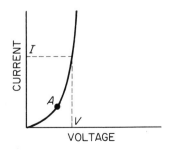

Figure 1-4. Diode Current–Voltage Curve

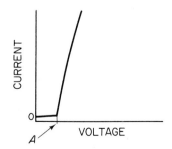

Figure 1-5. Approximate Current–Voltage Curve

### 1.1.4 Resistance Characteristics

*Resistance* for direct current and alternating current is another important characteristic of a junction diode. The resistance is, of course, closely related to current and voltage. For dc resistance, it is merely necessary to determine the current produced by some level of voltage, and then apply *Ohm's law:*

$$R_{dc} = \frac{V}{I}$$

For ac resistance, it is a matter of determining how much the current changes when the voltage is changed by a specified amount. These values are then used in the Ohm's law equation:

$$r_{ac} = \frac{\Delta V}{\Delta I}$$

The information for calculating both dc and ac resistance can be extracted from the current–voltage characteristics graph. Consider the curve in Figure 1-6. At the point where the voltage level is $V1$, the diode has a current of $I1$. The dc resistance at this point is

$$R_{dc} = \frac{V1}{I1}$$

At the point where $V2$ is the applied voltage, the dc resistance is

$$R_{dc} = \frac{V2}{I2}$$

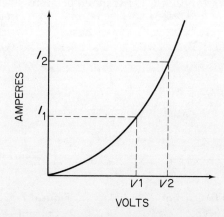

Figure 1-6.  Calculating Resistance

# Diodes and Thyristors

These two points produce almost identical values of dc resistance. The same is true of any point along the linear portion of the curve. Of course, if voltage–current values are taken in the lower, nonlinear portion of the curve, the dc resistance varies from one point to the next.

Using the same graph for ac resistance, the voltage change from $V1$ to $V2$ produces a current change from $I1$ to $I2$. The ac resistance then is

$$r_{ac} = \frac{\Delta V}{\Delta I}$$

$$= \frac{V2 - V1}{I2 - I1}$$

Voltage levels $V1$ and $V2$ may be peak-to-peak values of an applied sine wave. In this case, the actual resistance to the ac sine wave could be calculated in this fashion. Again there is an approximation for calculating ac resistance. The equation is

$$r_{ac} = \frac{26}{I_{av} \text{ (mA)}}$$

This is 26 divided by the average value of direct current. At room temperature, this equation produces approximately the same results for ac resistance as those obtained from the curves and Ohm's law. This approximation is accurate enough in general design work for application to all common junction-type diodes.

### 1.1.5 Equivalent Circuit

A diode can be represented by a circuit that includes the cut-in voltage and the ac resistance. Such an *equivalent circuit* is illustrated in Figure 1-7. The diode symbol is included to show that current can pass in only one direction. The battery, $V$, represents the cut-in voltage. Significant current will be present when the applied voltage, $V_a$, exceeds the cut-in voltage.

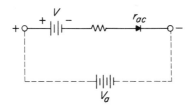

Figure 1-7.   Diode Equivalent Circuit

### 1.1.6 Leakage Current

It was stated earlier that leakage current can be ignored in most cases. This is true, but some discussion on the point is necessary because of the possibility of reverse current breakdown, which will destroy a common diode. When reverse bias is placed on a diode, *leakage current* (current from anode to cathode) will reach saturation in a few microseconds with a bias of about 0.1 V. Of course, this saturation current is measured in microamperes. A further increase in reverse bias voltage causes no appreciable change in leakage current until the breakdown point is reached. When this happens, the reverse current jumps to avalanche proportions and destroys the diode. These conditions are illustrated in Figure 1-8.

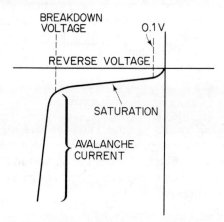

Figure 1-8.  Reverse-Bias Conditions

## 1.2  Zener Diode

Not all diodes are destroyed when subjected to reverse avalanche current. The Zener is such a diode. It is designed to operate in the avalanche current region. In fact, the avalanche current region is frequently referred to as the *Zener region*. Zeners are widely used as voltage reference diodes and for constant-current supplies. The Zener symbol is shown in Figure 1-9.

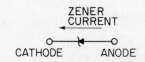

Figure 1-9.  Zener-Diode Symbol

## 1.2.1 Characteristics

The *Zener diode* is a silicon rectifier that is designed to operate in the breakdown region. Once breakdown voltage is reached, the voltage across the diode is held at a nearly constant value over a wide variation of current. The breakdown voltage, as well as current and power capabilities, are characteristics of individual Zeners. These characteristics are functions of the material and construction. The breakdown voltages, called *Zener voltages,* range from 1 volt to several hundred volts. The current and power ratings also vary widely, depending on the junction area and the method of cooling. Figure 1-8 is a fair description of the current–voltage characteristics of an operational Zener diode.

## 1.2.2 Biasing

The diagram in Figure 1-10, although not a practical circuit, illustrates the reaction of a Zener to both forward and reverse bias. The circuit has a 90-V Zener diode in series with a 15-kilohm (kΩ) resistor and the secondary of a transformer. The transformer applies an RMS (effective) ac voltage of 120 V. The peak values of the applied voltage are 170 V both ways from a zero reference. As the top of the transformer swings positive, the Zener is forward-biased and conducts like any other diode. The Zener is now shorted, and all the voltage appears across the resistor. As the top of the transformer swings negative, the Zener is reverse-biased. There is a negligible reverse current until the negative voltage reaches the breakdown potential of 90 V. As the sine wave drops from zero to −90 V, the Zener is open, and all the voltage appears across the Zener. At −90 V, the Zener goes into avalanche breakdown. As the voltage changes from −90 V to −170 V, Zener current increases, but the voltage across the Zener remains at 90 V. The increasing current drops more and more voltage across the resistor. Within the current limits, a Zener will maintain its voltage drop within about 1 V of the breakdown potential.

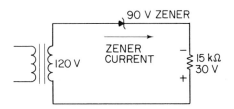

Figure 1-10. Zener Diode in a Series Circuit

## 1.3 Tunnel Diode

According to Ohm's law, the current through a resistance increases as the voltage across it increases:

$$I = \frac{E}{R}$$

A device that satisfies this requirement is said to have *positive resistance*. However, there are some devices that cause a decrease in current when voltage increases. These devices have a *negative resistance*. The *tunnel diode* is such a device. The symbol for a tunnel diode is illustrated in Figure 1-11.

Figure 1-11. Tunnel-Diode Symbol

### 1.3.1 Tunneling

Most semiconductors pass current only after sufficient voltage is applied to enable current carriers to cross the depletion zone and surmount the potential barrier of the junction. This potential is only a few tenths of a volt and was previously referred to as the cut-in voltage. The tunnel diode is a small *PN*-junction diode with a high concentration of impurities. The high impurity density forms a depletion zone so narrow that electrical charges can transfer across the junction long before they have enough energy to surmount the barrier. This transfer is a quantum-mechanical action called *tunneling*. The current carriers appear to tunnel through the potential barrier.

### 1.3.2 Current–Voltage Characteristics

Figure 1-12, the current–voltage characteristics graph, illustrates the tunneling action as well as the negative resistance of the tunnel diode. The figure indicates that tunneling begins the instant that the forward bias rises above zero. The current increases nearly perpendicularly as the forward bias is raised to about 0.06 V. Current then peaks at point A. As forward bias increases from 0.06 V to 0.4 V, the current decreases. As bias increases from 0.4 V to 0.8 V, the current rises again. The portion of the curve from zero to point A shows the tunneling action. Between points A and B, the diode has a negative resistance area. Beyond point B, the tunnel diode behaves as a conventional diode; current increases with the voltage.

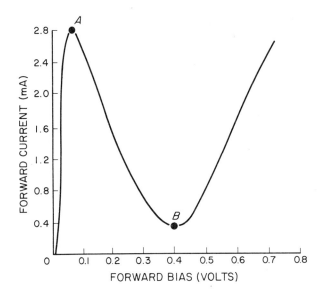

Figure 1-12.   Tunnel-Diode Characteristics

### 1.3.3   Operating Point

The tunnel diode may be used for amplifiers, oscillators, and switching circuits as well as for both forward and reverse rectification. The placement of the operating point is governed by the job that the diode is expected to perform.

The *operating point* for amplification should be approximately halfway between points A and B of Figure 1-12. This is in the center of the negative-resistance region and allows for a maximum current swing both ways from the operating point. In this case, the dc load line will be very steep so that it crosses the characteristic curve only once, as illustrated in Figure 1-13.

Noise is an important consideration in amplifiers, and in tunnel diodes it can be controlled somewhat by the placement of the operating point. For maximum amplification without distortion, the operating point should be in the center of the negative resistance region. For minimum noise, the operating point should be at a lower level of forward current. There must be a trade-off between maximum amplification and minimum noise.

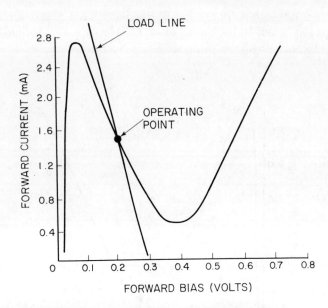

Figure 1-13. Tunnel-Amplifier Load Line

The tunnel diode can be used as a forward rectifier. This is seldom done because a conventional junction diode can handle this job as well or better than a tunnel diode. The forward voltage or positive resistance area of the characteristics curve is emphasized in Figure 1-14. If the tunnel diode is biased to operate in this region, it becomes a forward rectifier.

## 1.4 Tunnel Rectifier

A slight variation in the doping process and construction of a tunnel diode produces the *tunnel rectifier*. Although still classed as a tunnel diode, the characteristics, use, and symbol are different. The tunnel rectifier has a very small negative resistance region and a very low peak for the forward current. It is designed for use as a reverse rectifier, and in some cases, it is called a *back diode*. The characteristics curve and schematic symbol are illustrated in Figure 1-15. As the drawing indicates, a very small reverse bias produces a substantial reverse current through the tunnel rectifier. This characteristic makes it very useful for rectifying signals that are too small for the common junction diode to

Diodes and Thyristors 13

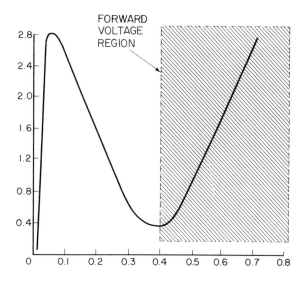

Figure 1-14. Forward-Voltage Area

handle. Tunnel rectifiers have a high-speed capability which, combined with their superior rectification properties, enables their use to couple signals in one direction while providing isolation in the other.

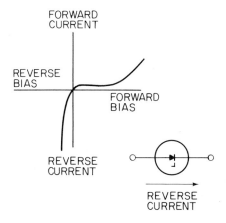

Figure 1-15. Tunnel Rectifier

## 1.5 Varactor Diode

The *varactor diode* is a *PN*-junction diode with a variable capacitance. This diode is designed to operate at microwave frequencies. The capacitance of the depletion region in the varactor diode has a nonlinear relation to the junction voltage. When operated with reverse bias, the varactor becomes the equivalent of a voltage-sensitive capacitor in series with a small value of resistance. This nonlinear capacitance and low series resistance enables the varactor to be used in oscillators, switching circuits, and frequency multipliers.

### 1.5.1 Construction

The varactor diode is constructed to have a very high impurity concentration outside the depletion region and a low concentration at the junction. This design feature causes the nonlinear capacitance by providing a capacitive reactance to the dominant current. The symbol for a varactor is shown in Figure 1-16.

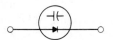

Figure 1-16.   Varactor Symbol

### 1.5.2 Characteristics

The current–voltage characteristics curve for a varactor is similar to that of other *PN*-junction diodes. The most important characteristic is the voltage–capacitance relationship, as illustrated in Figure 1-17. Notice that the junction capacitance is maximum when reverse voltage is minimum and decreases as the reverse voltage increases. The minimum capacitance is limited by the breakdown point, which must not be exceeded. The avalanche current would short the capacitance even if it did not destroy the diode. The maximum capacitance is limited by the point where voltage changes from reverse to forward bias. A midway point in the capacitance curve is designated as the control point. On this graph, the control voltage is 4 V, and this results in a capacitance of about 35 picofarads (pF).

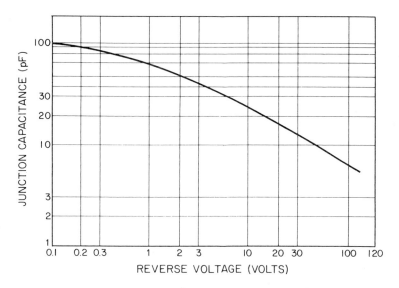

Figure 1-17.  Voltage–Capacitance Curve

An equation can be constructed from the curve that approximates the junction capacitance at any given level of voltage:

$$C_j = \frac{C_{jz}}{\sqrt[3]{1 + (0.167 \times V)}}$$

where $C_j$ is junction capacitance at the specified voltage, $C_{jz}$ the junction capacitance at zero bias, and $V$ the specified reverse voltage.

## 1.6  Silicon-Controlled Rectifier

The *silicon-controlled rectifier* (SCR) is not a diode and it is not a rectifier. The SCR is a thyristor. The silicon-controlled rectifier is frequently confused with the *silicon rectifier*. Actually, the silicon rectifier is a small junction diode, whereas the silicon-controlled rectifier is a four-layer three-junction electronic switch. This structure places the SCR in the thyristor family of semiconductors.

### 1.6.1  Structure

The structure and schematic symbol of an SCR are illustrated in Figure 1-18. The diagram shows a structure of *P-N-P-N* going from

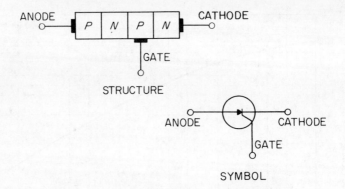

Figure 1-18. Silicon-Controlled Rectifier

anode to cathode. A device structured in this manner is operated by an anode potential more positive than the cathode. It is triggered to the on state by applying a potential, usually a pulse, to the gate, which makes the gate positive with respect to the cathode. The order of the layers could be reversed. In this case, the SCR becomes *N-P-N-P*, and the operating potentials would also have reversed polarities. However, the *P-N-P-N* is the most common order, so it will be used for the remainder of this discussion.

### 1.6.2 Characteristics

The principal voltage–current characteristics of the SCR are shown in Figure 1-19. The diagram shows that when the anode is positive with respect to the cathode, the SCR has an off state and an on state. At low values of anode-to-cathode voltage, the SCR has a very high impedance, which allows a very slight forward current called the *off-state current*. When this voltage is increased to the point of breakover voltage, the SCR switches to the on state. In the on state, the SCR is almost a dead short; the current is limited almost entirely by the impedance of the external circuit. It will remain in the on state as long as the external circuit draws a current in excess of the holding current. When the circuit current drops below the holding level, the SCR switches back to the off state.

A small reverse current is present when the anode is negative with respect to the cathode. This is called *reverse blocking current,* and it remains at a very low level until the voltage reaches the reverse breakdown point. When reverse breakdown voltage is applied, the SCR goes into a condition known as *thermal runaway.* During thermal runaway,

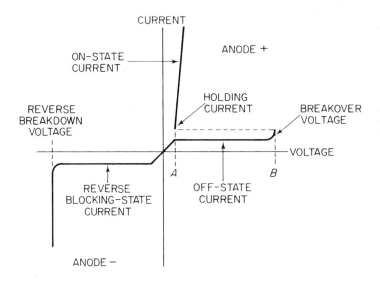

Figure 1-19. SCR Characteristics

the current is out of control, it increases very rapidly, and it usually destroys the SCR. The reverse breakdown voltage is generally about 100 times as great as the forward breakover voltage.

### 1.6.3 Gating Action

The breakover voltage point can be controlled anyplace between points A and B by applying a trigger pulse to the gate. The breakover voltage decreases as the amplitude of the gating pulse increases. In practice, the anode-to-cathode voltage is well below the breakover voltage point, and a gating pulse of sufficient magnitude is applied to assure that the SCR will switch on at the desired instant. Once the SCR is on, the gate has no more control. The SCR stays on as long as the external circuit draws enough current to sustain conduction. Sustaining current is any current in excess of the holding current. The light-sensitive SCR uses a beam of light to provide the gating pulse.

## 1.7 Triac

The word *triac* is an acronym that was coined from the term "triode ac switch." It is another three-electrode thyristor. This device is equivalent to two SCRs in parallel and oriented in opposite directions. The triac is controlled by a single gate, and it can be used as an on–off switch for current in either direction. In some cases, the triac is called a "bidirectional triode thyristor." The triac was developed by General Electric. It is not uncommon for triacs to handle current up to 25 amperes (A) and voltages to 500 V. Since it is an ac switching device for power frequencies, a triac can be obtained for either 60 hertz (Hz) or 400 Hz.

### 1.7.1 Structure

Since the triac has an on state and an off state in both directions, it must be constructed in a manner that allows each electrode to be in contact with both *P*- and *N*-type materials. This involves five or more junctions and very precise placement of the electrode contacts. This arrangement is illustrated in Figure 1-20. Notice that the main terminals

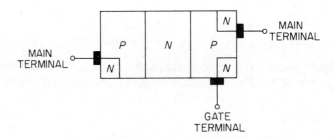

Figure 1-20. Triac Structure

are simply labeled 1 and 2 instead of cathode and anode. Either of these main terminals will act as an emitter in one direction and a collector in the opposite direction. And since the gate is in contact with both *P*- and *N*-type materials, a positive gate pulse triggers the triac on in one direction while a negative gate pulse triggers it on in the opposite direction.

Main terminal 1 is used as the reference point for all polarities. When terminal 2 is positive (in respect to terminal 1), the device is a *P-N-P-N* triac progressing from terminal 2 to terminal 1. When main terminal 2 is negative, the triac is still a *P-N-P-N* device, but now the order progresses from terminal 1 to terminal 2. In either condition, the triac can be triggered by either a positive or a negative gate pulse.

The triac symbol is shown in Figure 1-21. The terminal designators are not part of the symbol, but terminal 1 is always on the same side as the gate, and the gate is always the diagonal lead.

# Diodes and Thyristors

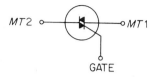

Figure 1-21. Triac Symbol

## 1.7.2 Characteristics

The triac has four triggering modes, two in quadrant I and two in quadrant III. A *triggering mode* refers to the relative potentials of the operating voltages. Taking all potentials in respect to main terminal 1, the four modes are as follows:

| MT 2 | Gate | Quadrant |
|---|---|---|
| positive | positive | I+ |
| positive | negative | I− |
| negative | positive | III+ |
| negative | negative | III− |

Notice that the polarity symbol following the quadrant number refers to the gate potential with respect to main terminal 1. Further operational characteristics are revealed in Figure 1-22.

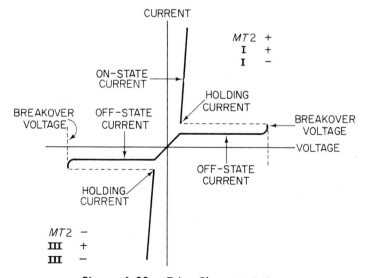

Figure 1-22. Triac Characteristics

The breakover voltage is the maximum level of direct voltage. When this level is reached (MT2 to MT1), the triac triggers itself into the on state. This breakover value must be greater than the peak alternating value applied between the main terminals to maintain control by the gate. When a trigger is applied to the gate in either quadrant, it must be of sufficient amplitude and duration to produce a *latching current*. The latching current is always slightly greater than the holding current. Once the triac is latched, it is fully on and will remain on until the current drops below the holding current level. There is also a maximum width for the triggering pulse. The maximum width is determined by the gate power rating, and it is affected by many factors. Care must be exercised that no maximum ratings are exceeded.

The most sensitive triggering mode is quadrant I+. The least sensitive mode is quadrant III+. For this reason, quadrant III+ should be avoided where it is possible to do so. When use of the III+ mode is necessary, a triac designed especially for this mode should be utilized.

### 1.7.3 Triggering

A triac may be triggered by any one of a variety of triggering circuits or devices. Triggering pulses are normally specified in terms of current. The current specifications are very important when the triggering circuit contains an inductor. In this case, the current lags the voltage. The voltage pulse must be applied sufficiently in advance to assure proper trigger current at the desired turn-on time. Probably the simplest method of triggering a triac is to apply an ac sine wave, as illustrated in Figure 1-23. The circuit being controlled by this triac would be connected in parallel to $R1$. When the triac is in the cutoff state, there is no voltage drop across $R1$. The off triac is an open circuit, and all the voltage is dropped across the open. When the triac is on, it is a short, and all the voltage appears across $R1$. The conducting angle is controlled by the resistance of $R2$. The output shows a conducting angle from 90 degrees of the input to 180 degrees and again

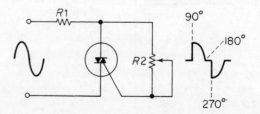

Figure 1-23. Triac Triggering

from 270 to 360 degrees. The resistor, R2, can be adjusted to turn on the triac before 90 degrees and before 270 degrees. Regardless of the turn-on point, this triac will conduct until the input sine wave passes through zero.

## 1.8 Diode and Thyristor Symbols

There are many symbols associated with diodes and thyristors, and most of them have become relatively standard. These symbols vary slightly with the different types of diodes. This section contains most of the common symbols in use. They are included here for ready reference. The common semiconductor symbols are included in Chapter Two.

### 1.8.1 Rectifier Symbols

| | |
|---|---|
| $C_s$ | shunt capacitance |
| $I_{FAV}$ | average forward current |
| $i_{FM(rep)}$ | peak repetitive forward current |
| $I_{RM}$ | maximum reverse current |
| $V_{RM}$ | maximum forward direct voltage drop |
| $V_{RM(nonrep)}$ | peak reverse voltage, nonrepetitive |
| $V_{RM(rep)}$ | peak reverse repetitive voltage |
| $V_{RMS}$ | effective (RMS) supply voltage |

### 1.8.2 Silicon-Controlled-Rectifier Symbols

| | |
|---|---|
| Critical $dv/dt$ | critical rate of applied forward voltage |
| $di_T/dt$ | rate of change of on-state voltage |
| $I_{DOM}$ | peak off-state current |
| $I_{GT}$ | average trigger current |
| $i_{HO}$ | ac holding |
| $I_{RRDM}$ | peak reverse current repetitive, open gate |
| $i_T$ | ac on state |
| $I_{T(AV)}$ | average on-state current |
| $I_{T(RMS)}$ | effective (RMS) on-state current |
| $I_{TSM}$ | nonrepetitive on-state surge current |
| $[I_{TS(RMS)}]^2 t$ | nonrepetitive effective (RMS) on-state surge current |
| $P_{G(AV)}$ | average power dissipation |

| | |
|---|---|
| $P_{GM}$ | peak gate power dissipation |
| $R_L$ | load resistance |
| $t_{gt}$ | gate-controlled turn-on time |
| $t_q$ | circuit-commutated turn-on time |
| $V_{DROM}$ | repetitive peak off-state voltage, open gate |
| $V_{DSOM}$ | nonrepetitive peak forward voltage, open gate |
| $V_{F(BO)O}$ | forward breakover voltage, open gate |
| $V_{GT}$ | average trigger voltage |
| $V_{RROM}$ | repetitive peak reverse voltage, open gate |
| $V_{RSOM}$ | nonrepetitive peak reverse voltage, open gate |
| $V_T$ | instantaneous on-state voltage |

### 1.8.3 Tunnel-Diode Symbols

| | |
|---|---|
| $C$ | total capacitance |
| $C_j$ | junction capacitance |
| $C_p$ | case capacitance |
| $C_{tv}$ | valley-point terminal capacitance |
| $f_{\max}$ | maximum frequency of oscillation |
| $f_{RO}$ | resistive cutoff frequency |
| $g_j$ | junction conductance |
| $I_F$ | forward current |
| $I_I$ | inflection-point current |
| $I_p$ | peak-point current |
| $I_P/C_{tv}$ | speed index |
| $I_v$ | valley-point current |
| $L$ | inductance, series |
| $L_{\text{ex}}$ | excessive inductance, series |
| $r_j$ | junction resistance |
| $r_s$ | series resistance |
| $t_{sw}$ | characteristic switching time |
| $V_F$ | forward voltage |
| $V_I$ | inflection-point voltage |
| $V_P$ | peak-point voltage |
| $V_{PP}$ | projected peak-point voltage |
| $V_R$ | reverse voltage |
| $V_v$ | valley-point voltage |
| $Y_t$ | terminal admittance |

### 1.8.4 Triac Symbols

| | |
|---|---|
| Commutating $dv/dt$ | rate of applied commutating voltage |
| Critical $dv/dt$ | critical rate of rise of off-state voltage |
| $I_{DROM}$ | peak off-state current |
| $I_{GT}$ | direct-current gate trigger |
| $I_{HO}$ | dc holding |
| $i_T$ | ac on state |
| $I_{T(RMS)}$ | effective (RMS) on-state current |
| $I_{TSM}$ | nonrepetitive peak surge, on-state current |
| $P_{G(AV)}$ | average gate power dissipation |
| $P_{GM}$ | peak gate power dissipation |
| $R_L$ | load resistor |
| $t_{gt}$ | gate-controlled turn-on time |
| $V_D$ | instantaneous off-state voltage |
| $V_{DROM}$ | repetitive peak off-state voltage |
| $V_{GT}$ | direct-voltage gate trigger |
| $V_T$ | instantaneous on-state voltage |

CHAPTER TWO

# Bipolar Transistors

A diode is a device formed by a single junction between $N$-type material and $P$-type material. A *bipolar transistor* is formed by making two such diodes in a single piece of material. A piece of $N$-type material sandwiched between two pieces of $P$-type material becomes *PNP* transistor. A piece of $P$-type material sandwiched between two pieces of $N$-type material becomes an *NPN* transistor. We shall discuss the basic structure, characteristics, functional circuits, and parameters of this three-layer two-junction device.

## 2.1 Structure

The transistor manufacturing process is not essential to this discussion. Suffice it to say that when a single piece of crystal contains two $PN$ junctions with proper terminals, it is a transistor. These junctions could be grown, alloyed, or diffused. The characteristics of these junctions *are* of primary concern.

### 2.1.1 PNP Transistor

As previously stated, and as indicated by the position of the letters, a *PNP transistor* has a piece of $N$-type material sandwiched between two pieces of $P$-type material. This basic structure is illustrated in Figure 2-1. Each of the three slabs of material has an assigned name. The center material is always the base, and it is much thinner than the two outer layers. The two outer materials are called *emitter* and *collector*. The emitter and collector are of the same type of material; the only difference is in the physical size. The collector is generally much larger

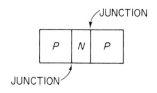

Figure 2-1.  Structure of a *PNP* Transistor

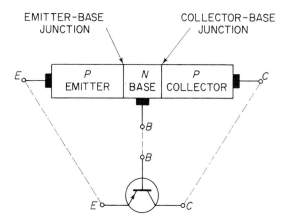

Figure 2-2.  Terminals and Symbol of a *PNP* Transistor

than the emitter. These names, terminals, and schematic symbol are illustrated in Figure 2-2.

The potential on the base controls the current between collector and emitter. The two junctions are the *emitter–base junction* and the *collector–base junction*. Each of these junctions has a depletion region and a barrier potential as described for diode junctions in Chapter One.

Notice in the schematic symbol that the emitter contains an arrow. In a *PNP* transistor, this arrow points toward the base. Electron current through a transistor always moves against the emitter arrow. Since this is a *PNP* transistor, the direction of current is *in* from both base and collector and *out* through the emitter.

### 2.1.2  NPN Transistor

The *NPN transistor* has a piece of *P*-type material sandwiched between two pieces of *N*-type material, as illustrated in Figure 2-3. In the *NPN* transistor, the center material is of the *P*-type. So it has a *P*-type base and *N*-type collector and emitter. The schematic symbol is illus-

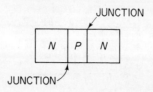

Figure 2-3. Structure of an NPN Transistor

trated in Figure 2-4. Compare this symbol with that in Figure 2-2. The only difference is the direction of the emitter arrow. In the *NPN* transistor, the arrow points outward from the base. Current direction through this transistor is *in* from the emitter and *out* from both base and collector.

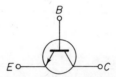

Figure 2-4. NPN Schematic Symbol

## 2.2 Controlling the Electron Flow

There is always some random movement of electrons inside a transistor. A free electron drops into a hole while another electron breaks loose and floats free. This type of random movement does not constitute current. In order to have a useful flow of electrons, external voltages must be applied that will not only direct the electron flow but will enhance the development of current carriers.

### 2.2.1 Basic Biasing Principles

A potential difference across a *PN* junction constitutes bias. When the voltage is connected with negative to the *N*-type material and positive to the *P*-type material, the junction has *forward bias*. When the polarity is reversed, positive to the *N*-type material and negative to the *P*-type material, the junction has *reverse bias*. In most applications, a transistor conducts *only* when the emitter–base junction is forward-biased and the collector–base junction is reverse-biased. The proper biasing polarities are illustrated in Figure 2-5. In this illustration, both

# Bipolar Transistors

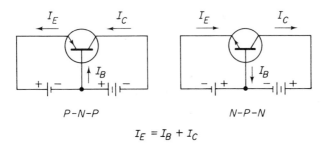

$$I_E = I_B + I_C$$

Figure 2-5. Proper Biasing Polarities

transistors have forward bias across the emitter–base junction and reverse bias across the collector–base junction. The emitter current, $I_E$, in both cases, is the total current. The base current, $I_B$, plus the collector current, $I_C$, is equal to the emitter current. The directions of the various currents are indicated by the polarities of the bias voltages. Electron current exists from negative to positive.

### 2.2.2 Distribution of Current

The amplification qualities of a transistor depend upon the distribution of current through the transistor. As previously stated, the emitter current is the total current. And as a general rule, the base current represents from 0.5 to 5 percent of this total current. The remaining 95 to 99.5 percent is the collector current. In most cases, the figure of 2 percent for base current and 98 percent for collector current is close enough for general discussion. Percentage and direction of currents for both types of transistor are illustrated in Figure 2-6.

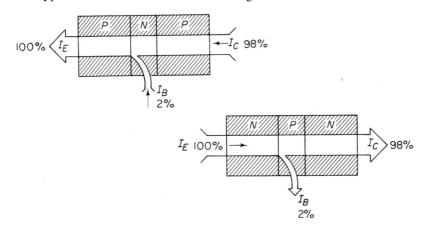

Figure 2-6. Percentage and Direction of Currents

In the *PNP*, 98 percent of the current enters through the collector; 2 percent enters through the base; and all the current emerges through the emitter. In the *NPN*, all the current enters through the emitter; 98 percent emerges through the collector; and 2 percent emerges through the base. These percentage values, of course, vary slightly from one transistor to another. In fact, the ratios among these currents are used to calculate the maximum gain of a transistor.

### 2.2.3 Current Ratios

There are two important current ratios to remember: the collector to emitter and the collector to base. The collector-to-emitter current ratio is call *alpha* and is symbolized by the lower case Greek letter of the same name ($\alpha$):

$$\alpha = \frac{I_C}{I_E} \quad \text{(with any constant value of collector voltage)}$$

These are direct currents, so alpha is the direct-current gain from emitter to collector. Since $I_C$ is always less than $I_E$, alpha is always less than unity.

The collector-to-base ratio is represented by the lower case Greek letter beta ($\beta$). Beta is the direct-current gain from base to collector:

$$\beta = \frac{I_C}{I_B} \quad \text{(with any constant value of collector voltage)}$$

The collector current, $I_C$, is always large with respect to the base current, $I_B$, so beta is always a relatively large number. Beta is usually about 50, but it may be as high as 1000.

Using the previous current percentages instead of measured current values, alpha and beta work out as follows:

$$\alpha = \frac{I_C}{I_E}$$
$$= \frac{98}{100}$$
$$= 0.98$$
$$\beta = \frac{I_C}{I_B}$$
$$= \frac{98}{2}$$
$$= 49$$

# Bipolar Transistors

When either alpha or beta is known, the other may be calculated. These are the relationships between the two:

$$\alpha = \frac{\beta}{\beta + 1}$$

$$\beta = \frac{\alpha}{1 - \alpha}$$

Even though these equations are for dc ratios, they are an indication of what can be expected. In fact, a slight alteration changes both alpha and beta to ac values. The ac alpha is

$$\alpha = \frac{\Delta I_C}{\Delta I_E} \quad \text{or} \quad \alpha = \frac{i_c}{i_e}$$

The ac beta is

$$\beta = \frac{\Delta I_C}{\Delta I_B} \quad \text{or} \quad \beta = \frac{i_c}{i_b}$$

The ac alpha is the forward current gain through a common base configuration. The ac beta is the forward current gain through a common emitter configuration. These configurations are covered in Section 2.3.

### 2.2.4 Leakage Current

There is another current that is always present in a transistor. This is leakage across the collector–base junction. *Leakage current* is labeled $I_{CBO}$, which means "current between collector and base with the emitter open." It is so designated because $I_{CBO}$ is measured with the emitter circuit open. This leakage current is opposite to the direction of normal base current but in the same direction as the collector current.

As temperature increases, $I_{CBO}$ doubles for each 8 to 10°C increase. Since $I_{CBO}$ adds to the collector current and subtracts from the base current, it reduces the control of $I_B$ over $I_C$. Excessive leakage current can cause excessive temperatures, which can build up to thermal runaway.

## 2.3 Basic Circuits

Bipolar transistors can be connected in six different ways, but there are only three basic circuit configurations that provide amplification. This discussion covers these three amplification configurations. These circuits are named according to the transistor element that is common to both input and output. They are *common base, common emitter,* and *common collector.* Since the common element is often connected to ground, the term *grounded* is sometimes substituted for common. Each of the three configurations has its own individual characteristics which make it especially suited for certain jobs.

### 2.3.1 Common Base

As the name indicates, the transistor in the *common-base* configuration has its base connected as the element that is common to both input and output circuits. Both *PNP* and *NPN* transistors can be used in this circuit; the only difference is in the polarities of the bias voltages. Figure 2-7 illustrates the circuit for each type of transistor. Notice the polarities

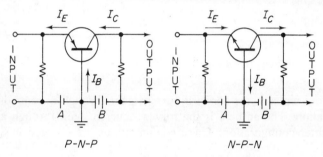

Figure 2-7.  Common-Base Circuits

of batteries A and B in both of these circuits. In both circuits, battery A is connected across the emitter–base junction with the negative terminal to the *N*-type material and the positive terminal to the *P*-type material. This polarity is forward bias, which is proper for the input junction. Electron current in the input section has a direction indicated by the battery. This causes current against the emitter arrow. Minimum operating bias for the emitter–base junction is approximately 0.3 V for a germanium transistor. The silicon transistor requires a higher input bias, approximately 0.7 V. In switching operations, the emitter–base junction may become reverse-biased. If this happens, care must be exercised to keep the reverse bias below 5 V. Reverse input bias in excess of 5 V will destroy most transistors.

Battery B in both circuits is connected with the negative terminal to the *P*-type material and the positive terminal to the *N*-type material. This polarity is reverse bias across the collector–base junction. In most cases, the output voltage supply is 20 V or less. However, this is *not* a rule, and much higher voltages can be used. In switching operations, the collector–base junction may also have its bias voltage polarities reversed. Such reversal places forward bias across the output junction. Such forward bias should be limited to about 0.5 V.

Remember that the emitter current, $I_E$, is always larger than the collector current, $I_C$. In this common-base configuration, the emitter is in the input circuit. So a change in input current causes a smaller change in output current. This input–output current ratio makes it impossible for the common-base configuration to produce a current gain larger than unity. However, it is a good amplifier for both voltage and power.

The input resistance is the opposition to current through the emitter–base junction. The output resistance is the opposition to current through the collector–base junction. Here are some sample values that may be expected from a common-base amplifier:

1. Input resistance: 30 to 150 Ω.
2. Output resistance: 300 to 500 kΩ.
3. Voltage gain: 300 to 1500.
4. Current gain: less than unity.
5. Power gain: 20 to 30 decibels (dB) (roughly 60 to 500 dB).

Complete current–voltage characteristics may be obtained from the static characteristics curves that are supplied by the manufacturer. A family of these curves for a particular transistor when used as a common-base amplifier is shown in Figure 2-8. This graph is a plot of emitter current and collector current for all possible values of collector voltage. Since the collector voltages are indicated as positive values, it may be assumed that the transistor is an *NPN*. However, the same graph applies equally well to a *PNP* transistor. If the transistor is indeed a *PNP*, all collector voltage values will be interpreted as being negative. The static curves only show the direct-current and direct-voltage conditions. A load line on this graph converts it into a dynamic characteristics graph. Load lines are discussed in Chapter Six.

### 2.3.2 Common Emitter

The *common-emitter* circuit is a very popular amplifier circuit because it amplifies voltage, current, and power. Figure 2-9 illustrates the common-emitter circuit arrangement for both *PNP* and *NPN* transistors. Notice in both cases that the emitter is common to both the input and the output circuit. The current gain in the common-emitter circuit is measured between base and collector. Since a very small change in base current produces a very large change in collector current, this configura-

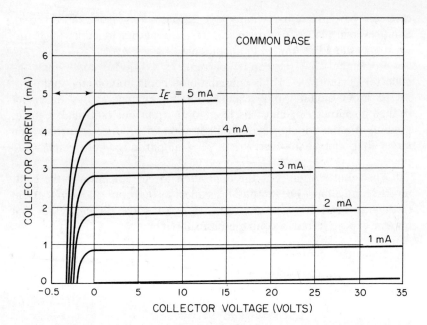

Figure 2-8. Static Characteristics for a Common Base

tion always has a current gain larger than unity. A current gain of 50 or more is common. Large voltage gains are also common, and this results in a very large power gain. Power gains in excess of 40 dB (over 10,000) can be realized with this circuit.

The direction of input current is indicated in both circuits by battery A; battery B shows the direction of the output current. In all cases, this electron current is against the emitter arrow. Notice that the currents through $R_B$ and $R_L$ are in opposite directions. An increase in $I_B$ causes

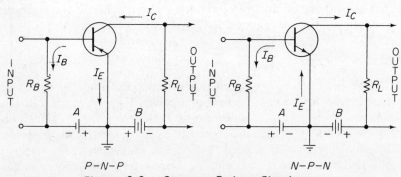

Figure 2-9. Common-Emitter Circuits

an increase in $I_C$, but the voltage across $R_B$ is 180 degrees out of phase with the voltage across $R_L$.

The input resistance is the resistance of the emitter–base junction, and in this circuit, that is on the order of 500 to 1500 Ω. The resistor from base to ground, $R_B$, parallels this emitter–base junction resistance, but the resistance of $R_B$ is kept large enough that it has negligible effect on the input resistance. The output resistance of the transistor is a combination of the resistances of both junctions. This output resistance ranges from 30 to 50,000 Ω. The output resistance of the circuit is modified by the value of $R_L$, which parallels the output resistance of the transistor.

Here are some sample values from common-emitter amplifiers:

1. Input resistance: 500 to 1500 Ω.
2. Output resistance: 30 Ω to 50 kΩ.
3. Voltage gain: 300 to 1000.
4. Current gain: 25 to 50.
5. Power gain: 25 to 40 dB (roughly 200 to 10,000 dB).

The manufacturer also provides static characteristics graphs for common-emitter amplifiers. A family of curves that composes such a graph reflects base current and collector current for all possible values of collector voltage. Such a graph is shown in Figure 2-10. This figure,

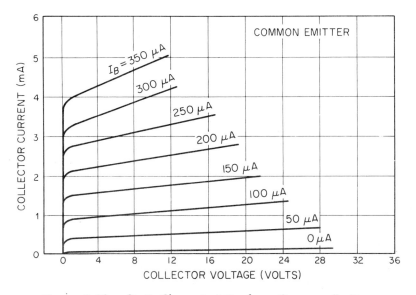

Figure 2-10. Static Characteristics for a Common Emitter

indicates positive values of collector voltage, $V_C$, and positive $V_C$ is correct for an *NPN* transistor. For a *PNP* transistor, all values of $V_C$ are negative. The graph differs slightly from one transistor to another, so the graph must match the transistor. Load lines can be constructed on this static graph to reveal how a transistor will perform in a given common-emitter circuit. Refer to Chapter Six for details on load-line construction.

### 2.3.3 Common Collector

Although the *common collector* always has a voltage gain less than unity, it is a very popular amplifier for other reasons. It does provide a gain of both current and power, and it is most valuable as an impedance-matching device. The high input resistance and low output resistance make it ideally suited for matching two diverse impedances. The common element in this circuit is the collector. The input is inserted across the collector–base junction, and the output is taken across the emitter load resistor. Basic circuits for both transistor types are illustrated in Figure 2-11.

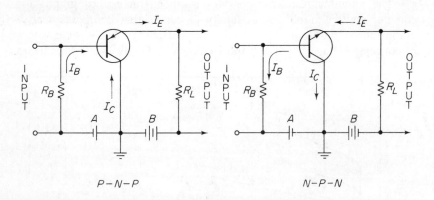

Figure 2-11. Common-Collector Circuits

Bias battery A in each case shows the direction of base current, $I_B$, while battery B shows the direction of collector current, $I_C$. The emitter current, as always, is the combined $I_B + I_C$, and its direction is against the emitter arrow. The resistance across the collector–base junction is on the order of 20,000 to 50,000 Ω. Resistor $R_B$ parallels this junction resistance and modifies the input resistance of the circuit. If minimal modification is desired, the value of $R_B$ should be at least 10

# Bipolar Transistors

times as large as the junction resistance. The circuit input resistance can be decreased by using smaller values of $R_B$.

The output resistance of the transistor is measured from emitter to collector. This resistance is a low value, between 50 and 1000 Ω. The emitter load resistor, $R_L$, parallels the transistor output resistance. The maximum output resistance for the circuit is obtained when the value of $R_L$ is at least 10 times the value of the transistor output resistance. The circuit output resistance can be reduced by using smaller values for $R_L$. The voltage changes across $R_B$ and $R_L$ are in phase with each other.

There are no characteristic curves specifically designed for the common collector. However, the common-emitter characteristics graph contains most of the essential information (see Figure 2-10). Current $I_B$ is the input current, and $I_E$ is the output current. Current gain is

$$\frac{\Delta I_E}{\Delta I_B}$$

The emitter current, $I_E$, is not given in Figure 2-10, but $I_E$ is always $I_C + I_B$. The graph does contain both $I_C$ and $I_B$. For example, suppose that with a $V_C$ of 16 V, a common collector has an $I_B$ of 150 microamperes (μA). The graph shows that this combination produces an $I_C$ of 1.8 mA:

$$I_E = I_C + I_B$$
$$= 1.8 \text{ mA} + 0.15 \text{ mA}$$
$$= 1.95 \text{ mA}$$

Increasing $I_B$ to 250 μA increases $I_C$ to 3.5 mA:

$$\Delta I_E = \Delta I_C + \Delta I_B$$
$$= 1.7 \text{ mA} + 0.1 \text{ mA}$$
$$= 1.8 \text{ mA}$$

Current gain then is

$$\frac{\Delta I_E}{\Delta I_B} = \frac{1.8 \text{ mA}}{0.1 \text{ mA}} = 18$$

The current gain is always approximately equal to $\beta + 1$. Since voltage gain is always slightly less than unity, the power gain is slightly less than the current gain.

Some sample values from common-collector amplifiers are as follows:

1. Input resistance: 20 to 500 kΩ.
2. Output resistance: 50 to 1000 Ω.
3. Voltage gain: less than unity.
4. Current gain: 25 to 50.
5. Power gain: 14 to 17 dB (roughly the same as current gain).

## 2.4 Practical Biasing Arrangements

Batteries were used in the basic circuits to illustrate the bias polarities, current directions, and the differences between *NPN* and *PNP* circuits. Batteries are not always practical for transistor bias, for several obvious reasons. It is time to drop the batteries and work toward more practical biasing arrangements. It should be clear now that the only difference between *NPN* and *PNP* transistor circuits is the polarity of the biasing voltages. So the dual-circuit presentation will also be dropped at this point.

The bipolar transistor is basically a current-operated device. Its operating point is established by the quiescent values of collector voltage and emitter current. The *quiescent state* is the condition that exists when there is no signal input. The current in the emitter–base circuit controls the current in the collector circuit. The particular requirements of the application and the transistor characteristics dictate what the operating point must be. The voltage and current requirements must be considered when selecting a biasing arrangement.

### 2.4.1 Common Base

A simplified biasing arrangement for a common-base amplifier is illustrated in Figure 2-12. A single dc power source provides the bias for both input and output circuits. The supply voltage, $V_{CC}$, is divided across $R1$ and $R2$. This places a positive potential on the base of the transistor. The relative values of $R1$ and $R2$ are selected to produce the desired level of emitter current, $I_E$. This emitter current through $R_E$ establishes a positive potential at the emitter but leaves the emitter negative with respect to the base. This relative potential across the emitter–base junction constitutes forward bias for this *NPN* transistor. The emitter current establishes the level of collector current. The collector current, $I_C$, through $R_L$ produces a voltage drop. The collector voltage

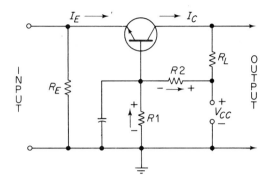

Figure 2-12. Common-Base Bias

is $V_{CC}$ minus the voltage across $R_L$. The base current, $I_B$, goes through $R2$ and back to the positive terminal of the voltage source. The capacitor is a stabilizer; it bypasses the ac signal around $R1$. So the values of $R1$ and $R2$ establish the base potential. The base potential determines the emitter current, and the emitter current determines the collector current.

### 2.4.2 Common Emitter

The circuit in Figure 2-13 illustrates a method of biasing a common-emitter amplifier. This biasing arrangement represents a trade-off between stability and gain. A small value of $R_E$ allows maximum voltage gain at a cost of instability. Increasing the value of $R_E$ makes the circuit more stable at a cost of reduced gain. For acceptable values of both gain and stability, $R_L$ should be about 10 times the value of $R_E$.

The $+V_{CC}$ makes the collector more positive than both base and emitter. The small $I_B$ through $R_B$ causes the base to be less positive than the collector but more positive than the emitter. This constitutes

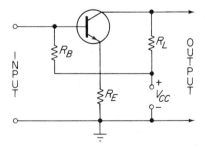

Figure 2-13. Common-Emitter Bias

forward bias across the emitter–base junction and reverse bias across the collector–base junction. The value of $R_B$ must be such that it establishes the proper value of bias across the emitter–base junction. This value should be about 0.5 V for a silicon transistor. At the operating point,

$$0.5 \text{ V} = I_B R_B - I_E R_E$$

### 2.4.3 Common Collector

Various types of voltage dividers can be used to supply bias voltages for the common collector. Speaking of an *NPN* transistor, the most positive element will be the collector, and the most negative element will be the emitter. The base must then be less positive than the collector and more positive than the emitter. For the *PNP* transistor, of course, all these polarities are just the opposite.

## 2.5 Transistor Tests

The designer of bipolar transistor circuits is concerned with five types of transistor tests. These tests are:

1. Collector leakage current.
2. Direct-current gain.
3. Voltage punch-through.
4. Alternating-current gain.
5. Four-terminal parameter test.

In the following circuits, standard *PNP* transistors are used, but the same tests can be used with *NPN* transistors by reversing the voltage polarities. The test circuits are derived from the equivalent circuits in a transistor tester.

### 2.5.1 Collector Leakage Current

There are two types of collector leakage current that increase with an increase in temperature. One leakage current is between the collector and base, and it is measured with the emitter open. This type of leakage is designated as $I_{CBO}$. Leakage $I_{CBO}$ decreases the base current and increases the collector current. The other leakage current is between collector and emitter, and it is measured with the base open. This second type of collector leakage is designated as $I_{CEO}$. Leakage current $I_{CEO}$ also increases the collector current. Excessive leakage current can cause thermal runaway and permanent damage to the transistor. On the other hand, a damaged transistor is very likely to have excessive leakage current. At any rate, leakage in excess of the specified maximum is an indication of a marginal or faulty transistor. The test circuit for $I_{CBO}$ is illustrated in Figure 2-14.

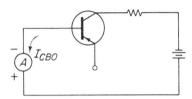

Figure 2-14. $I_{CBO}$ Test Circuit

As far as the base is concerned, $I_{CBO}$ is a reverse current. The ammeter for measuring this current is in the micro range. The battery is connected in the same manner as $V_{CC}$ would be connected in a normal circuit, and it provides normal reverse bias across the collector–base junction.

The test circuit for leakage current $I_{CEO}$ is illustrated in Figure 2-15. This leakage current is much larger than $I_{CBO}$, but it is not a reverse current. Current $I_{CEO}$ increases both the emitter current and the collector current. It should be kept in mind that both $I_{CBO}$ and $I_{CEO}$ are series aiding with the collector current. If temperature rises, leakage current increases; and when leakage current increases, the junction temperature rises and causes the leakage current to increase some more. If the transistor is expected to operate at other than room temperature, 25°C, the temperature derating curve must be considered.

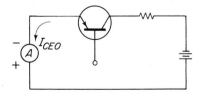

Figure 2-15. $I_{CEO}$ Test Circuit

### 2.5.2 Direct-Current Gain

The dc gain test measures the effectiveness of the base in controlling the collector current. This test is useful for a common-emitter configuration for most low-frequency applications. The test circuit is illustrated in Figure 2-16. Resistor $R1$ is adjusted for a null on the voltmeter. At this time, the dc gain is

$$\frac{I_C}{I_B}$$

This gain can also be determined by the ratio of $R_x$ to $R2$. The dc gain is

$$\frac{R_x}{R2}$$

Resistor $R_x$ is part of $R1$, and $R1$ generally has a calibrated dial that indicates the dc gain directly without the need for calculations. This type of gain is sometimes called *alpha dc*.

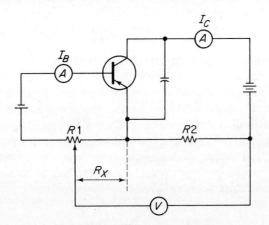

Figure 2-16. Dc-Gain-Test Circuit

### 2.5.3 Alternating-Current Gain

Alternating-current gain depends on the type of circuit configuration. In a common-base circuit, ac gain is the ratio of change between $I_C$ and $I_E$ with any constant value of collector voltage. Alternating-current gain for this configuration is always less than unity, and it is commonly referred to as *alpha:*

$$\alpha = \frac{\Delta I_C}{\Delta I_E}$$

When the circuit is a common emitter, ac gain is called beta ac. *Beta ac* is the ratio of change between $I_C$ and $I_B$ with any constant value of collector voltage:

$$\beta = \frac{\Delta I_C}{\Delta I_B}$$

The test circuits for both alpha and beta are similar to that in Figure 2-16. The differences are:

1. Insert an ammeter in the emitter circuit.
2. Apply an ac signal between the base and the emitter.
3. Remove the shorting capacitor.

This provides a common-emitter circuit. With a null on the voltmeter, the values of $I_C$ and $I_B$ will vary with the ac input. The change in $I_C$ divided by the change in $I_B$ will reveal the value of beta. Once either alpha or beta is known, it is a simple operation to calculate the other.

$$\alpha = \frac{\beta}{1 + \beta} \quad \text{and} \quad \beta = \frac{\alpha}{1 - \alpha}$$

When the transistor is to be placed in a common-collector circuit, the ac gain will be just slightly more than beta. In fact, $\beta + 1$ is acceptable for ac gain in a common collector. The term $\beta + 1$ is the ratio of a change in emitter current to a change in base current with $V_C$ constant. Values for both alpha and beta are indicated on the transistor data sheet. The measured and calculated values for alpha and beta should match the given values pretty closely.

### 2.5.4 Punch-Through Voltage

The punch-through-voltage test is used to determine the quality of a transistor that is to be used in a switching circuit. *Punch-through voltage* is the potential between collector and base that will cause a low-resistance path between collector and emitter. This is the point where the base region virtually disappears and the transistor action stops. Conducted with reasonable care, this test does not damage the transistor. The test circuit is illustrated in Figure 2-17. The variable dc is adjusted until punch-through voltage is obtained. Punch-through voltage is indicated when $M1$ reaches a specified reference level. At this time, the value of punch-through voltage is indicated on $M2$. The two limiting resistors keep $I_E$ and $I_C$ to safe values and prevent transistor damage. The presence of punch-through voltage should be limited to a very short time. The meters can be read in a few seconds at most.

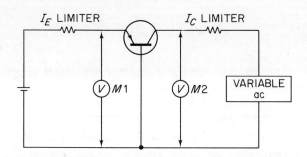

Figure 2-17. Punch-Through-Voltage Test Circuit

### 2.5.5 Four-Terminal Parameter

In general design work, the transistor may be considered as a four-terminal network to determine the relationships of inputs and outputs. These relationships are called *hybrid (h) parameters*. These $h$ parameters are referred to in data sheets and on test instruments. The four-terminal network is represented in Figure 2-18. In this arrangement,

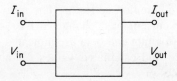

Figure 2-18. Four-Terminal Network

there are two currents and two voltages to be concerned with. If the two currents are considered as dependent variables, the resulting parameters are short-circuit parameters, and they are measured in mhos. When the two voltages are considered as dependent variables, the resulting parameters are open-circuit parameters, and they are measured in ohms. The four hybrid parameters are obtained by using one current and one voltage as dependent variables. The four hybrid parameters are:

$h_i$ = input $Z$ with output shorted

$h_r$ = reverse $V$ ratio with input open

$h_f$ = forward $I$ gain with output shorted

$h_o$ = output admittance with input open

The unit of measure for $h_i$ is the ohm and for $h_o$ is the mho. There is no unit for either $h_r$ or $h_f$ since they are transfer ratios.

These $h$ parameters can be applied to any of the three basic amplifier configurations. An additional subscript letter is added to each parameter to identify the basic configuration that the parameter pertains to. These subscripts are $b$ for common base, $e$ for common emitter, and $c$ for common collector. Alpha in a common-base circuit is $h_{fb}$, while beta in a common-emitter circuit is $h_{fe}$. A simplified test circuit for measuring the $h$ parameters is illustrated in Figure 2-19. Generator $G1$ is a calibrated current generator, and $G2$ is a calibrated voltage generator. Meter M measures the alternating current indirectly. Switches A, B, C, and D are ganged on a single rotor. With the switch in position 1, the calibrated current generator ($G1$) is connected to the emit-

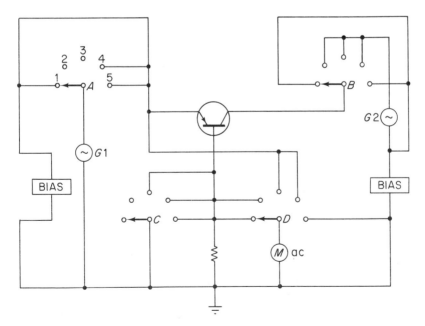

Figure 2-19. Hybrid-Parameter Test Circuit

ter of the transistor while the collector is grounded. Meter $M$ now indicates the ratio between the ac in the emitter and the base: $i_e/i_b$. This ratio is equivalent to $h_{fe} + 1$ (or beta $+ 1$), which is the forward current gain for a common collector.

In position 2 of the switch, the calibrated voltage generator ($G2$) is connected to the collector and the emitter is open. Meter $M$ is indicating the output admittance ($h_o$). In position 3, the meter indicates $h_r$; in position 4, it indicates $h_i$; and in position 5, it indicates alpha ($h_f$).

Refer to Chapter Five for additional information on parameters.

## 2.6 Common Symbols

There are many symbols used in solid-state manuals and on bipolar transistor circuits. Most books briefly explain each symbol when it is first encountered. This section is provided as a handy reference when an unfamiliar symbol is encountered.

### 2.6.1 General Solid-State Symbols

| | |
|---|---|
| df | duty factor |
| $\eta$ (eta) | efficiency |
| NF | noise figure |
| $T$ | temperature |
| $T_A$ | ambient temperature |
| $T_C$ | case temperature |
| $T_J$ | junction temperature |
| $T_{MF}$ | mounting flange temperature |
| $T_S$ | soldering temperature |
| $T_{STG}$ | storage temperature |
| $\Theta$ | thermal resistance |
| $\Theta_{JA}$ | thermal resistance, junction to ambient |
| $\Theta_{JC}$ | thermal resistance, junction to case |
| $\Theta_{JHS}$ | thermal resistance, junction to heat sink |
| $\Theta_{JMF}$ | thermal resistance, junction to mounting flange |
| $t$ | time |
| $t_d$ | time delay |
| $t_d + t_r$ | turn-on time |
| $t_f$ | fall time |
| $t_p$ | pulse time |

# Bipolar Transistors

| | |
|---|---|
| $t_r$ | rise time |
| $t_s$ | storage time |
| $t_s + t_f$ | turn-off time |
| $\tau$ (tau) | time constant |
| $\tau_s$ | time constant, saturation stored charge |

## 2.6.2 Bipolar Transistor Symbols

| | |
|---|---|
| $C_c$ | capacitance, collector to base |
| $C_{cb}$ | capacitance, feedback collector to base |
| $C_{ibo}$ | capacitance, common-base input, open circuit |
| $C_{ieo}$ | capacitance, common-emitter input, open circuit |
| CM | cross modulation |
| $C_{obo}$ | capacitance, common-base output, open circuit |
| $C_{oso}$ | capacitance, common-emitter output, open circuit |
| $E_{s/b}$ | energy cutoff frequency, second breakdown |
| $f_c$ | cutoff frequency |
| $f_{hfb}$ | small-signal forward-current transfer ratio cutoff frequency, common base, short circuit |
| $f_{hfe}$ | small-signal forward-current transfer ratio cutoff frequency, common emitter, short circuit |
| $f_T$ | gain-bandwidth product, frequency where $h_f$ equals unity |
| $g_{me}$ | small-signal transconductance, common emitter |
| $G_{PB}$ | large-signal average power gain, common base |
| $G_{pb}$ | small-signal average power gain, common base |
| $G_{PE}$ | large-signal average power gain, common emitter |
| $G_{pe}$ | small-signal average power gain, common emitter |
| $h_{FB}$ | static forward-current transfer ratio, common base |
| $h_{fb}$ | small-signal forward-current transfer ratio, common base |
| $h_{FE}$ | static forward-current transfer ratio, common emitter |
| $h_{fe}$ | small-signal forward-current transfer ratio, common emitter |
| $h_{ib}$ | small-signal input impedance, common base, short circuit |
| $h_{IE}$ | static input resistance, common emitter |

| | |
|---|---|
| $h_{ie}$ | small-signal input inpedance, common emitter, short circuit |
| $h_{ob}$ | small-signal output impedance, common base, open circuit |
| $h_{oe}$ | small-signal output impedance, common emitter, open circuit |
| $h_{rb}$ | small-signal reverse-voltage transfer ratio, common base, open circuit |
| $h_{re}$ | small-signal reverse-voltage transfer ratio, common emitter, open circuit |
| $I_B$ | dc in base |
| $i_b$ | ac in base |
| $I_{b1}$ | turn-on current |
| $I_{b2}$ | turn-off current |
| $I_C$ | dc in collector |
| $i_c$ | ac in collector |
| $I_{CB}$ | collector cutoff current |
| $I_{CBO}$ | leakage current between collector and base with emitter open |
| $I_{CEO}$ | leakage current between collector and emitter with base open |
| $I_{CER}$ | current between collector and emitter with specified resistance between base and emitter |
| $I_{CES}$ | current between collector and emitter with base shorted to emitter |
| $I_{CEV}$ | current between collector and emitter with specified voltage between base and emitter |
| $I_{CEX}$ | current between collector and emitter with specified circuit between base and emitter |
| $I_{CS}$ | switching current at minimum $h_{FE}$ |
| $I_E$ | dc in emitter |
| $i_e$ | ac in emitter |
| $I_{EBO}$ | current between emitter and base with collector open |
| $I_{c/b}$ | collector current, second breakdown |
| MAG | maximum-available amplifier gain |
| MUG | maximum-usable amplifier gain |
| $P_{BE}$ | average dc power input to base, common emitter |

| | |
|---|---|
| $p_{be}$ | ac power input to base, common emitter |
| $P_{CB}$ | dc power output at collector, common base |
| $p_{cb}$ | ac power output at collector, common base |
| $P_{CE}$ | dc power output at collector, common emitter |
| $p_{ce}$ | ac power output at collector, common emitter |
| $P_{EB}$ | dc power input to emitter, common base |
| $p_{eb}$ | ac power input to emitter, common base |
| $P_{IB}$ | large-signal input power, common base |
| $p_{ib}$ | small-signal input power, common base |
| $P_{IE}$ | large-signal input power, common emitter |
| $p_{ie}$ | small-signal input power, common emitter |
| $P_{OB}$ | large-signal output power, common base |
| $p_{ob}$ | small-signal output power, common base |
| $P_{OE}$ | large-signal output power, common emitter |
| $p_{oe}$ | small-signal output power, common emitter |
| $Q_s$ | stored base charge |
| $r_{bb}'$ | intrinsic base spreading resistance |
| $r_b' C_c$ | collector-to-base time constant |
| $r_{CE(\text{sat})}$ | saturation resistance, collector to emitter |
| $R_e$ | same as $h_{ie}$, real part of small-signal input impedance, common emitter, short circuit |
| $R_G$ | generator resistance |
| $R_{ie}$ | input resistance, common emitter |
| $R_L$ | load resistance |
| $R_{oe}$ | output resistance, common emitter |
| $R_s$ | source resistance |
| $\tau$ (thermal) | thermal time constant |
| $V_B$ | direct base voltage |
| $V_{BB}$ | base supply voltage |
| $v_b$ | alternating base voltage |
| $V_{BC}$ | direct base-to-collector voltage |
| $V_{BE}$ | direct base-to-emitter voltage |
| $V_{BE(\text{sat})}$ | base-to-emitter saturation voltage |
| $V_{(BR)CBO}$ | collector-to-base breakdown voltage, emitter open |
| $V_{(BR)CEO}$ | collector-to-emitter breakdown voltage, base open |
| $V_{(BR)CER}$ | collector-to-emitter breakdown voltage, specified resistance between base and emitter |

| | |
|---|---|
| $V_{(BR)CES}$ | collector-to-emitter breakdown voltage, base shorted to emitter |
| $V_{(BR)CEV}$ | collector-to-emitter breakdown voltage, specified voltage between base and emitter |
| $V_{(BR)EBO}$ | emitter-to-base breakdown voltage, collector open |
| $V_{CB}$ | direct collector-to-base voltage |
| $v_{cb}$ | alternating collector-to-base voltage |
| $V_{CB(fl)}$ | floating potential collector to base, emitter-base bias, collector–base open |
| $V_{CBO}$ | collector-to-base voltage, emitter open |
| $V_{CBV}$ | collector-to-base voltage, specified voltage between emitter and base |
| $V_C$ | direct collector voltage |
| $V_{CC}$ | collector supply voltage |
| $v_c$ | alternating collector voltage |
| $V_{CEO}$ | collector-to-emitter voltage, base open |
| $V_{CER}$ | collector-to-emitter voltage, specified resistance between base and emitter |
| $V_{CES}$ | collector-to-emitter voltage, base shorted to emitter |
| $V_{CEV}$ | collector-to-emitter voltage, specified voltage between base and emitter |
| $V_{CE(\text{sat})}$ | collector-to-emitter saturation voltage |
| $V_{EB}$ | emitter-to-base voltage |
| $V_{EBO}$ | emitter-to-base voltage, collector open |
| $V_E$ | direct emitter voltage |
| $V_{EE}$ | emitter supply voltage |
| $v_e$ | alternating emitter voltage |
| $V_{RT}$ | reach-through voltage |
| **VG** | voltage gain |
| $1/Y_{22(\text{real})}$ | real part of short-circuit impedance |
| $Y_{fe}$ | forward transconductance |
| $Y_{ie}$ | input admittance |
| $Y_{oe}$ | output admittance |
| $Y_{re}$ | reverse transconductance |

**CHAPTER THREE**

# Field-Effect Transistors

*Field-effect transistors* are unipolar, voltage-sensitive devices that use an electrostatic field to control conduction. They combine the inherent advantages of small size, low power, and mechanical ruggedness with a high input impedance. In addition, they have a square-law transfer characteristic that makes them ideally suited for low cross modulation. They are in many ways superior to bipolar transistors, although both types still have a definite place in electronic design. There are two general types of field-effect transistors: junction (JFET) and insulated gate (IGFET). The operation is similar in the two types, but the structure is different.

## 3.1 Junction Field-Effect Transistor

The *junction field-effect transistor* (JFET) is a single-junction device that relies on only one type of current carrier. There are two types of JFETs: the *N*-channel and the *P*-channel. The *N*-channel conducts by means of electron carriers. The *P*-channel uses hole carriers. The two types are constructed and operated in like manner. The differences are in the arrangement of the *P*- and *N*-type materials and the polarity of the operating voltages.

### 3.1.1 Structure of an N-channel JFET

The *N-channel JFET* is constructed by diffusing a circle of *P*-type material around a piece of *N*-type material. This circle forms one continuous *PN* junction. The *N*-type material then becomes a channel for electron current carriers, and the *P*-type material is the controlling element. Three leads are attached: one to each end of the channel and one to the controlling element. The leads are called *source, drain,* and *gate.* The structure is illustrated in Figure 3-1. The figure shows that the doughnut-like ring of *P*-type material has been diffused into the *N*-type material. The source and drain, unless labeled otherwise, can be interchanged. The electron current originates at the source and is collected by the drain.

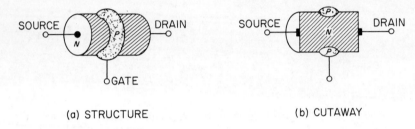

(a) STRUCTURE    (b) CUTAWAY

Figure 3-1. Structure and Biasing of an N-channel JFET

The two most used symbols for an *N*-channel JFET are shown in Figure 3-2. The arrow is now on the gate element, and it points to the *N*-type channel. The only difference between symbols a and b is the position of the arrow. The symbol used in any specific instance is a matter of individual choice.

Figure 3-3 shows the proper voltage polarities and the effect of different bias levels. A relatively large positive potential is connected between the drain and the source. This voltage is designated as $V_{DS}$ (voltage drain to source). A smaller negative potential, $V_{GS}$, is connected between the gate and the source. This arrangement places the drain as the most positive element and the gate as the most negative element.

Increasing the negative potential on the gate widens the depletion

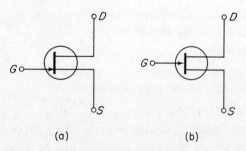

(a)    (b)

Figure 3-2. Schematic Symbols for an N-channel JFET

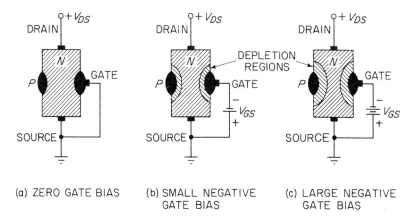

(a) ZERO GATE BIAS    (b) SMALL NEGATIVE GATE BIAS    (c) LARGE NEGATIVE GATE BIAS

Figure 3-3. Biasing of an N-channel JFET

region around the junction and decreases the current (parts b and c). The gate control of current is complete: all the way from saturation to cutoff.

### 3.1.2 Characteristics of an N-channel JFET with Zero Gate Bias

In this configuration, the gate is connected directly to the source as shown in part a of Figure 3-3. The drain current–voltage characteristics for zero gate bias ($V_{GS} = 0$) are shown on the graph of Figure 3-4. When the drain-to-source voltage, $V_{DS}$, is zero, the drain current is zero. As $V_{DS}$ increases, $I_D$ starts and increases in a linear fashion until the current nears saturation. Now the drain current increase slows down until pinch-off voltage, $V_p$, is reached. When $V_{DS}$ reaches the $V_p$ point,

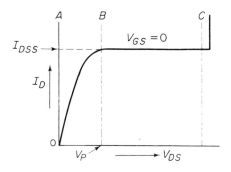

Figure 3-4. Drain Characteristics ($V_{GS} = 0$)

the drain current flattens out. This value of drain current is known as the *drain–source saturation current,* or $I_{DSS}$. This $I_{DSS}$ is normally considered to be the maximum $I_D$ with the gate shorted to the source. This is the same as saying that $V_{GS} = 0$. When $V_{DS}$ is increased beyond the $V_p$ point, there is little or no change in drain current. However, if $V_{DS}$ is increased to the breakdown point, the JFET goes into avalanche conduction with a sudden large increase in drain current.

The area on the graph from point A to point B is the ohmic channel or triode region. The reason for these names is the fact that the channel is behaving like a resistor, similar to the action of a triode tube. The area from point B to point C is the pinch-off region. For all values of $V_{DS}$ in the pinch-off region, the drain current remains at the saturation level. The JFET should be operated at $V_{DS}$ levels well below the breakdown potential. When driven to breakdown, the avalanche effect may destroy the transistor. Normal operating values of $V_{DS}$ fall anywhere within the pinch-off region.

### 3.1.3 Characteristics of an N-channel JFET with Gate Bias

A family of current–voltage characteristics curves is produced by plotting a separate curve for each of several values of gate voltage. The graph in Figure 3-5 is such a family of curves. Now it can be seen that the controlling factor is the potential applied to the gate. For each different level of $V_{GS}$, the areas on the graph change slightly. Maximum $I_D$ occurs at different values of $V_{GS}$, pinch-off voltage changes, and the point of breakdown changes. Notice that both $V_p$ and breakdown po-

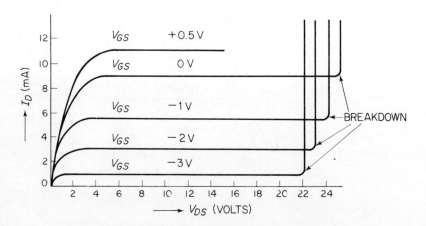

Figure 3-5. Drain Characteristics with Gate Bias

# Field-Effect Transistors

tential occur at lower values of $V_{DS}$ as the gate potential becomes more negative. Such a family of curves presents only the (static) direct current and voltage characteristics. Load lines can be constructed on these graphs to indicate dynamic conditions in a particular circuit. See Chapter Seven for a detailed discussion on the use and construction of JFET load lines.

### 3.1.4 Transfer Characteristics of an N-channel JFET

The transfer characteristics graph presents a direct relationship between input voltage, $V_{GS}$, and output current, $I_D$. Dynamic transfer graphs are generally available, but if not, they are easily constructed from a current–voltage characteristics graph. The relationship between the two graphs is shown in Figure 3-6. The transfer graph is drawn to the left of the current–voltage characteristics graph.

This graph is simply a locus of all points of maximum drain current and the gate voltages at those points. The current-bias points are located at some constant value of $V_{DS}$. Again, these are dc characteristics, but now there is a direct relation between $V_{GS}$ and $I_D$. With this graph, the effect of input voltage on the output current is readily apparent.

The following equation is an approximation of the JFET transfer characteristics:

$$I_D = I_{DSS} \left(1 - \frac{V_{GS}}{V_p}\right)^2$$

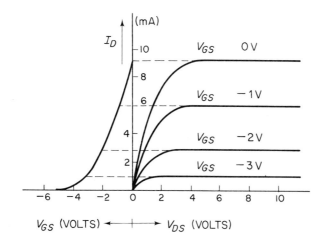

Figure 3-6. Construction of a Transfer Characteristics Graph

For example: when $I_{DSS}$ is 9 mA, $V_{GS}$ is 1 V, and $V_p$ is 2 V, then:

$$I_D = 9 \text{ mA} \left(1 - \frac{1}{2}\right)^2$$
$$= 9 \text{ mA}(0.5)^2$$
$$= 9 \text{ mA} \times 0.25$$
$$= 2.25 \text{ mA}$$

### 3.1.5 Structure of a P-channel JFET

The *P-channel JFET* is constructed by diffusing a ring of *N*-type material around a piece of *P*-type material as illustrated in Figure 3-7.

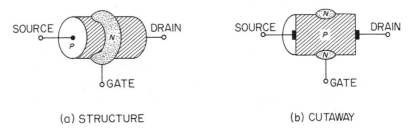

(a) STRUCTURE  (b) CUTAWAY

Figure 3-7. Structure of a P-channel JFET

Now the channel is composed of *P*-type material and the majority current carriers are the holes. The controlling element, the gate, is of *N*-type material. This reversal of material causes a reversal of all biasing potentials. The direction of electron movement is from drain to source, so $V_{DS}$ is a negative potential. The gate-to-source junction must have reverse bias, so the $V_{GS}$ is a positive potential.

The two most used schematic symbols for the *P*-channel JFET are exactly the same as those for the *N*-channel JFET except for the direction of the gate arrow. The arrow now points away from the *P*-type channel. Bias and symbols are illustrated in Figure 3-8.

### 3.1.6 Characteristics of a P-channel JFET

There are only two differences between *N*-channel characteristics and *P*-channel characteristics: the current direction and voltage polarities are reversed. Compare the graph in Figure 3-9 with that in Figure 3-5; then compare Figure 3-10 with Figure 3-6.

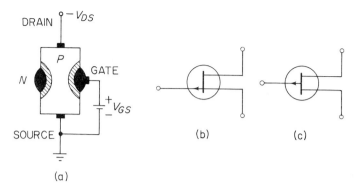

Figure 3-8.  Bias and Schematic Symbols for a P-channel JFET

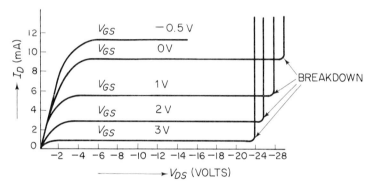

Figure 3-9.  Drain Characteristics of a P-channel JFET with Gate Bias

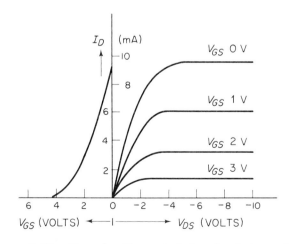

Figure 3-10.  Transfer Characteristics of a P-channel JFET

### 3.1.7 JFET Transconductance

*Transconductance* and *forward transfer admittance* are two names for the same characteristic. Transconductance is designated by $g_m$, and $\tau_{fs}$ indicates forward transfer admittance. Regardless of the name used, it is a measure of the effectiveness of gate voltage in controlling the drain current. Test measurements are taken with $V_{DS}$ held to a constant value. The equation is

$$g_m \text{ (or } \tau_{fs}) = \frac{\Delta I_D}{\Delta V_{GS}} \quad \text{(with } V_{DS} \text{ constant)}$$

The unit of measure is mho, and the deltas specify a small change. When $I_D$ and $V_{GS}$ are in amperes and volts, respectively, the result is in mhos, the basic unit. However, $g_m$ is such a small quantity that it is usually expressed in micromhos.

The levels of $I_D$ and $V_{GS}$ for calculating $g_m$ are readily available on the transfer characteristics graph. Consider the graph in Figure 3-11.

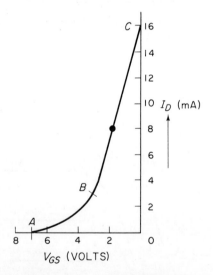

Figure 3-11. Calculating $g_m$

An operating point may be selected to correspond to any point along the transfer curve. The information for calculating transconductance is obtained by making a small change in $V_{GS}$ and noting the resulting change in $I_D$. On the curved portion of the characteristic, point A to point B, $g_m$ changes from point to point. On the linear portion of the

# Field-Effect Transistors

characteristic, point B to point C, the value of $g_m$ holds fairly constant. At the point where $V_{GS} = 2$ V and $I_D = 8$ mA, a change to $V_{GS} = 1$ V causes $I_D$ to increase to 11 mA. The $g_m$ in this area is

$$g_m = \frac{\Delta I_D}{\Delta V_{GS}}$$

$$= \frac{3 \text{ mA}}{1 \text{ V}}$$

$$= 3 \text{ mmhos} \quad \text{or} \quad 3000 \text{ } \mu\text{mhos}$$

### 3.1.8 Drain Resistance of a JFET

The resistance between drain and source is different for ac and dc. The dc resistance is designated as either $R_{DS}$ or $R_{D(on)}$, and it symbolizes the dc resistance of the channel when the depletion region is removed. This, of course, is the zero gate bias condition. The ac resistance between drain and source is designated as $r_d$ or $r_{DSS}$. This ac resistance is the dynamic output resistance of the JFET. The equation for ac resistance is

$$r_d = \frac{\Delta V_{GS}}{\Delta I_D} \quad \text{(with } V_{DS} \text{ constant)}$$

Instead of ac resistance, the data sheet may have the output admittance designator, $\tau_{os}$. Since admittance is the reciprocal of resistance, it is a simple matter to change from one to the other:

$$\tau_{os} = \frac{1}{r_d}$$

Resistance is measured in ohms; admittance is measured in mhos.

### 3.1.9 Input Resistance of a JFET

The very high input resistance is one of the many strong features of the JFET. This input resistance is the resistance across the reverse-biased gate–channel junction. The designator symbol for input resistance is $R_{GS}$. This resistance, $R_{GS}$, is generally on the order of $10^9$ Ω at room temperature. This resistance drops slightly as temperature rises. At 100°C, the value of $R_{GS}$ ranges around $10^7$ Ω, which is still a substantial resistance.

A reverse, minority charge current exists across the reverse-biased junction. This current is designated as $I_{GSS}$, which means gate-to-source current with the drain shorted to the source. Current $I_{GSS}$ is commonly referred to as either gate reverse current or gate-to-source cutoff current. At 25°C, $I_{GSS}$ is on the order of 1 nanoampere (nA), and it increases as temperature rises. This increase in $I_{GSS}$ is inevitable because of the decrease in the input resistance.

## 3.2 Insulated Gate Field-Effect Transistor

*Insulated gate* is a descriptive name for a type of FET that actually has no junction between the channel and the gate. In fact, the conducting channel between source and drain (in one type) must be induced by the potential on the gate. This insulated gate FET (IGFET) relies heavily on metal oxide in the construction process, and for this reason, it is commonly referred to as MOSFET, which means *metal oxide semiconductor field-effect transistor*. Two of the important differences between JFET and MOSFET are as follows:

1. The extremely high input resistance of the MOSFET is unaffected by the polarity of the gate potential.
2. The leakage current in a MOSFET is largely independent of the temperature.

The MOSFET is particularly well suited for such applications as RF amplifiers, voltage control circuits, and voltage amplifiers. There are two general types of MOSFET: the channel-enhancement and the channel-depletion types.

### 3.2.1 Structure of a Channel-Enhancement MOSFET

The *channel-enhancement MOSFET* takes its name from the fact that the conduction channel between source and drain is nonexistent until the proper bias is applied. The channel must be induced by applying forward bias between gate and source. The enhancement-channel MOSFET is divided into the $N$-channel and the $P$-channel types. The $N$-channel type is discussed here, but reversing the $P$- and $N$-type materials will cause the information to apply to the $P$-channel type.

The *N-channel MOSFET* begins with a lightly doped $P$-type substrate of highly resistant semiconductor material. This substrate is usually in the form of a thin silicon wafer. The wafer is polished, then oxidized in a furnace with a thin layer of insulating metal oxide. Spots are then etched through the oxide layer for the source and the drain. The wafer now goes into a furnace with a gas of $N$-type material. The

# Field-Effect Transistors

*N*-type material diffuses into the two spots of exposed *P*-type substrate. Another oxidation process insulates the areas of exposed silicon, then small spots are etched for attaching the source and drain contacts. Metal is now evaporated over the entire wafer surface. Another etching process removes the excess metal and separates the drain, source, and gate contacts.

Connector wires are bonded to the metal contacts to form the electrode leads. Each unit is hermetically sealed in a case in an inert atmosphere. The transistor is then tested, and all external leads are shorted together. These leads should remain shorted together until the MOSFET is soldered into its circuit. The short is intended to prevent static charges from damaging the gate insulation.

The drawing in Figure 3-12 is greatly exaggerated to illustrate the mechanical structure. Notice that there is no conduction channel between the two spots of *N*-type material that constitute source and drain.

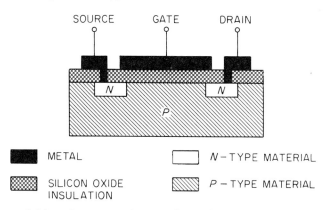

Figure 3-12. Structure of an *N*-channel Enhancement MOSFET

This conduction channel must be induced by placing a positive potential on the gate. This positive will attract electrons in the substrate and cause them to collect in the area between source and drain. The concentration of electrons in this area has the same effect as a strip of *N*-type material. The gate-positive potential has enhanced the negative carriers and created a conduction channel. In the absence of a positive potential on the gate, there can be no useful current between source and drain. Even with zero gate bias, there is no current. So the enhancement MOSFET is normally off, and the gate bias turns it on.

The schematic symbols for the enhancement MOSFET are illustrated in Figure 3-13. The lead with the arrow is the substrate. It is sometimes labeled with a U and sometimes with a B. If there is such an external lead, it should be connected to the circuit ground. If this fourth

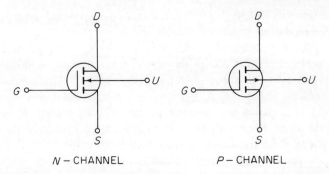

Figure 3-13.   Schematic Symbols for an Enhancement MOSFET

external lead is not present, it means that the substrate is internally connected to the source. The direction of the substrate arrow indicates the type of channel. The arrow points toward the $N$-type channel and away from the $P$-type channel. The broken line between source and drain identifies the enhancement MOSFET.

### 3.2.2   Characteristics of an Enhancement MOSFET

The graph in Figure 3-14 represents the drain and transfer characteristics of an $N$-channel enhancement MOSFET. The $N$-channel type is implied by the fact that all voltages on the graph are positive. The same graph with negative voltages represents a $P$-channel enhancement MOSFET.

This graph shows that the conductivity of the channel is enhanced by the positive potential on the gate. Notice that the drain current increases as the $V_{GS}$ becomes more positive. The fact that the gate is in-

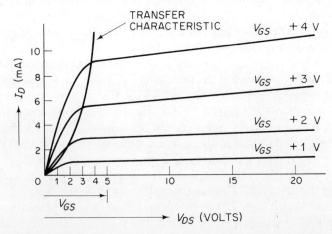

Figure 3-14.   Characteristics of an Enhancement MOSFET

sulated from the channel accounts for the extremely high input resistance and the virtual absence of leakage current.

### 3.2.3 Structure of a Channel-Depletion MOSFET

Again, the discussion is centered on the *N*-channel type but applies equally well to the *P*-channel type when the material types are reversed. The structure of the *channel-depletion MOSFET* is identical to that described for the enhancement MOSFET, with one exception. The depletion MOSFET has a lightly doped *N*-type conduction channel diffused between the source and the drain. This structure is illustrated in Figure 3-15. The channel-depletion MOSFET has current from source to drain with zero bias on the gate. In fact, it requires a considerable reverse bias to deplete the electron carriers in the channel and cut it off. The channel-depletion MOSFET is normally on.

The schematic symbols for the channel-depletion MOSFET are illustrated in Figure 3-16. Notice that these symbols have a solid bar

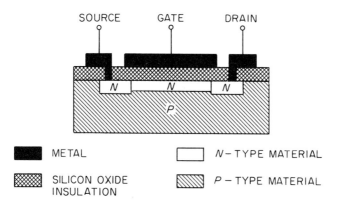

Figure 3-15. Structure of a Channel-Depletion MOSFET

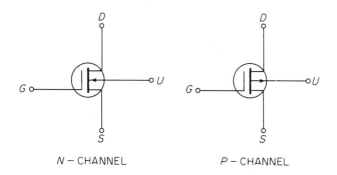

Figure 3-16. Schematic Symbols for a Depletion MOSFET

representing the conduction channel between source and drain. The arrow still points in for the *N*-channel type and out for the *P*-channel type. The lead marked U is the substrate, and it is handled as explained under enhancement-channel MOSFET. The gate is separated from the channel as before. This signifies that the gate is insulated.

### 3.2.4 Characteristics of a Channel-Depletion MOSFET

The depletion-channel MOSFET can be operated with the gate either positive, negative, or zero with respect to the source. When operated with zero bias on the gate, the diffused strip of doped material between source and drain functions reasonably well as a conduction channel. Applying a positive to the gate sets up a field that attracts more electron carriers into the channel area. These electrons enhance the *N*-channel carriers, and the drain current increases as the positive gate potential increases. When the gate is made negative, the resulting field attracts positive (holes) carriers into the channel area while repelling the negative electron carriers. This action depletes the *N*-channel carriers, and the drain current decreases as the gate becomes more negative. So the resistance of the channel depends upon the polarity and level of the applied gate voltage. The characteristics are summed up by saying that the depletion-channel MOSFET has two modes of operation: the depletion mode and the enhancement mode. When operated with a positive on the gate, it is in the enhancement mode. When it has a negative on the gate, it is in the depletion mode. In fact, this transistor is often called the *depletion–enhancement MOSFET*. The two modes are illustrated in Figure 3-17.

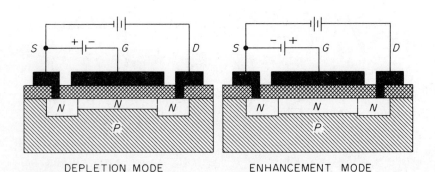

Figure 3-17. Modes of Operation

Field-Effect Transistors

As always, the drain is positive with respect to the source. In the depletion mode, $V_{GS}$ is negative, and this depletes the carriers in the $N$-type channel. There is less conduction because the channel resistance has gone up. When $-V_{GS}$ reaches approximately 4 V, the transistor will cut off. This complete stoppage of drain current indicates that for all practical purposes, the channel has ceased to exist.

In the enhancement mode, $V_{GS}$ is positive, and this attracts electrons from the substrate to increase the channel area. The $N$-type channel is enhanced, and drain current is increased. The characteristics are further illustrated in Figure 3-18. This figure shows that in the depletion mode,

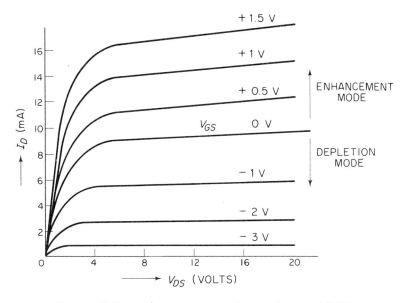

Figure 3-18. Characteristics of a Depletion MOSFET

$I_D$ is largely independent of the value of $V_{DS}$. The only requirement here is that $V_{DS}$ be some positive potential. In the enhancement mode, $I_D$ increases slightly with an increase of $V_{DS}$. This graph is for an $N$-channel depletion MOSFET. For the $P$-channel type, the polarity of $V_{DS}$ becomes negative and the $V_{GS}$ polarities are reversed. These differences are shown in Figure 3-19. Since the gate potential may be either positive or negative, the transfer characteristics for the $N$-channel depletion MOSFET are approximately as depicted in the graph of Figure 3-20.

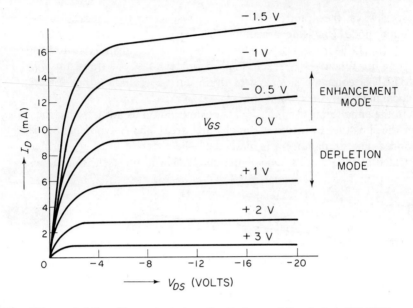

Figure 3-19.   Characteristics of a P-channel Depletion MOSFET

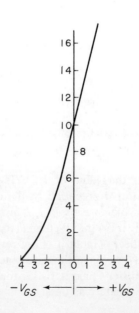

Figure 3-20.   Transfer Characteristics of a Depletion MOSFET

## 3.3 Basic FET Circuits

The FET is also used in three basic circuit configurations. These are common gate, common source, and common drain. The name is taken from the element that is placed in a position to be *common* to both input and output circuits. Since this common element is generally grounded, the word *grounded* is sometimes used instead of common. The common-gate configuration finds popular application in high-frequency voltage amplifiers. The common-drain configuration provides high power gain and makes a good buffer amplifier. The common-source configuration lends itself to a variety of uses because it amplifies voltage, current, and power. The common source is used more often than the other two. The information in this section applies to all types of field-effect transistors, but with slight reservations.

### 3.3.1 Common-Gate Circuit

The *common-gate circuit,* as the name implies, has the gate as the element that is common to both input and output. This configuration is illustrated in Figure 3-21. This is an $N$-channel enhancement MOSFET.

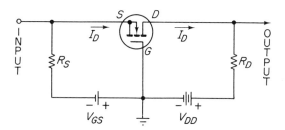

Figure 3-21. Common Gate

The substrate arrow and the voltage polarities are reversed for the $P$-channel type. In this arrangement, the common element is readily apparent. The gate-to-source voltage places the gate slightly positive with respect to the source. The drain supply voltage places the drain positive with respect to both source and gate, with the source being the most negative. The positive $V_{GS}$ attracts electrons from the substrate to create an $N$-type channel between source and drain. The input signal is applied across $R_S$ and the output is taken across the drain load resistor, $R_D$.

The common gate effectively has an input impedance equal to the value of $R_S$. Since $R_S$ is in series with $R_D$, the drain current $I_D$ passes through both $R_S$ and $R_D$. The value of $R_D$ must be at least 10 times as large as that of $R_S$ to realize a reasonable gain. Even with a large ohmic value for $R_S$, this is considered a low input impedance for a MOSFET. This characteristic makes the circuit useful for matching a low impedance to a high impedance.

The common gate has a low voltage amplification, which may be considered a disadvantage of the circuit. However, the low gain has some advantage in circuit design because it allows the neutralization circuits to be eliminated in many applications.

The following equation may be used to calculate the voltage gain of a common gate circuit:

$$A_v = \frac{(g_m r_d + 1) R_D}{(g_m r_d + 1) R_S + r_d + R_D}$$

where $g_m$ is forward transconductance, $r_d$ the output resistance of the device, $R_S$ the source resistance, and $R_D$ the output load resistance.

The following values are not necessarily typical, but they are fairly close to what might be expected from a MOSFET.

$$g_m = 2000 \; \mu\text{mhos}$$
$$r_d = 8000 \; \Omega$$
$$R_S = 600 \; \Omega$$
$$R_D = 2000 \; \Omega$$

These values may be substituted into the gain equation to solve for voltage gain:

$$\begin{aligned} A_v &= \frac{(g_m r_d + 1) R_D}{(g_m r_d + 1) R_S + r_d + R_D} \\ &= \frac{(2000 \times 10^{-6} \times 8000 + 1)2000}{(2000 \times 10^{-6} \times 8000 + 1)600 + 8000 + 2500} \\ &= \frac{322 \times 10^3}{107 \times 10^3} \\ &= 3 \end{aligned}$$

There is another equation that is somewhat less cumbersome, and it renders approximately the same results. It is

$$A_v = \frac{g_m R_D}{g_m R_S + 1}$$

# Field-Effect Transistors

where $R_D$ is the drain resistance and $R_S$ is the source resistance. Using the values from the previous example enables a comparison of the results:

$$A_v = \frac{g_m R_D}{g_m R_S + 1}$$

$$= \frac{2000 \times 10^{-6} \times 2000}{2000 \times 10^{-6} \times 600 + 1}$$

$$= \frac{40}{13}$$

$$= 3$$

Generally, the input impedance is approximately the same as the source resistance $R_S$. The equation is

$$Z_i = \frac{g_m R_S + 1}{g_m}$$

In previous examples $R_S$ was 600 Ω and $g_m$ was 2000 μmhos. So the input impedance must have been

$$Z_i = \frac{g_m R_S + 1}{g_m}$$

$$= \frac{2000 \times 10^{-6} \times 600 + 1}{2000 \times 10^{-6}}$$

$$= \frac{1300}{2}$$

$$= 650 \, \Omega$$

The output impedance is approximately equal to the value of the drain resistor $R_D$:

$$Z_o = R_D$$

However, $r_d$, the output resistance of the device, is in parallel with $R_D$ and may modify the output impedance of the circuit. In this case, $Z_o$ is calculated by

$$Z_o = \frac{r_d R_D}{r_d + R_D}$$

The values previously used for $r_d$ and $R_D$ indicate that $r_d$ must be considered: $r_d = 8000\ \Omega$ and $R_D = 2000\ \Omega$. Using these in the equation,

$$Z_o = \frac{r_d R_D}{r_d + R_D}$$

$$= \frac{8 \times 10^3 \times 2 \times 10^3}{8 \times 10^3 + 2 \times 10^3}$$

$$= \frac{16 \times 10^6}{10 \times 10^3}$$

$$= 1.6\ k\Omega$$

In this instance, $r_d$ lowered the output impedance from $2000\ \Omega$ to $1600\ \Omega$. If $r_d$ is less than 10 times the value of $R_D$, the impedance modification must be taken into consideration.

### 3.3.2 Common-Source Circuit

The *common-source configuration* is the most popular MOSFET circuit because it provides good voltage amplification together with reasonable gain in both current and power. An N-channel enhancement MOSFET is used in Figure 3-22 to illustrate the common-source circuit.

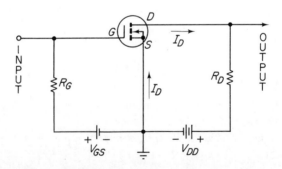

Figure 3-22. Common Source

The grounded source is the most negative point in the circuit, and the drain is the most positive. The small voltage, $V_{GS}$, keeps the gate slightly more positive than the source. The input is applied across $R_G$ to the gate. The output is taken across the drain load resistor $R_D$.

The common-source circuit combines a very high input impedance with a reasonably high output impedance. The value of $R_G$ is on the

## Field-Effect Transistors

order of 1 million $\Omega$, which gives the circuit a high input impedance. The output impedance of the device is high enough that it has little effect on the circuit output impedance. The circuit output impedance is then approximately equal to the value of $R_D$. The value of $R_D$ has a direct bearing on the voltage gain; the voltage gain is directly proportional to the value of $R_D$.

The voltage gain for the common source is

$$A_v = \frac{g_m r_d R_D}{r_d + R_D}$$

Assume that $g_m$ is 5000 $\mu$mhos, $r_d$ is 8000 $\Omega$, and $R_D$ is 10,000 $\Omega$. The voltage gain then becomes

$$A_v = \frac{g_m r_d R_D}{r_d + R_D}$$

$$= \frac{5 \times 10^{-3} \times 8 \times 10^3 \times 10 \times 10^3}{8 \times 10^3 + 10 \times 10^3}$$

$$= \frac{400 \times 10^3}{18 \times 10^3}$$

$$= 22.2$$

In this example, $r_d$ has a considerable effect on the gain because there is little difference in the values of $r_d$ and $R_D$. In most cases, $r_d$ has a much larger value. When the value of $r_d$ is at least 10 times the value of $R_D$, the voltage gain becomes

$$A_v = g_m R_D$$

Using figures from the previous example,

$$A_v = g_m R_D$$

$$= 5000 \times 10^{-6} \times 10 \times 10^3$$

$$= 50$$

The common source does have a voltage phase inversion. As the signal on the gate swings positive, the voltage on the drain swings negative, and vice versa.

At low frequencies, the input and output impedances are unaffected by capacitance. So the low-frequency input impedance for the circuit is

$R_G$, in parallel with the internal gate–source resistance, $R_{GS}$. For all practical applications, $R_{GS}$ is an open circuit, so

$$Z_i = R_G$$

When the frequencies are high enough to cause capacitance to become a significant factor, then total input $X_C$ appears in parallel with $R_G$. In this case, input impedance becomes

$$Z_i = \frac{R_G X_C}{R_G + X_C}$$

The output impedance for low frequencies is approximately the same as $R_D$:

$$Z_o = R_D$$

However, this changes as frequencies increase. The output capacitance is the drain-to-source capacitance. When this becomes significant, the output impedance becomes

$$Z_o = \frac{R_D X_C}{R_D + X_C}$$

### 3.3.3 Common-Drain Circuit

The *common-drain circuit* has a voltage gain less than unity, but it is frequently used as a power amplifier. An N-channel enhancement MOSFET is used in Figure 3-23 to illustrate the common-drain configuration. Even though the drain is grounded, it is still the most positive point

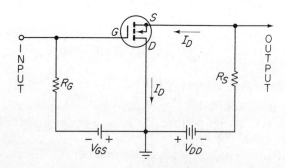

Figure 3-23. Common Drain

# Field-Effect Transistors

in the circuit. The small voltage, $V_{GD}$, keeps the gate potential less positive than the drain but still more positive than the source. The input is applied across $R_G$ to the gate. The output is taken across $R_S$.

The exact voltage gain is calculated by

$$A_v = \frac{g_m r_d R_S}{r_d + R_S + g_m r_d R_S}$$

Assuming a $g_m$ of 3000 μmhos, an $r_d$ of 100 kΩ, and an $R_S$ of 10 kΩ produces these results:

$$\begin{aligned}
A_v &= \frac{g_m r_d R_S}{r_d + R_S + g_m r_d R_S} \\
&= \frac{3 \times 10^{-3} \times 100 \times 10^3 \times 10 \times 10^3}{100 \times 10^3 + 10 \times 10^3 + (3 \times 10^{-3} \times 100 \times 10^3 \times 10 \times 10^3)} \\
&= \frac{3000 \times 10^3}{3110 \times 10^3} \\
&= 0.965
\end{aligned}$$

Notice, in this case, that $r_d$ is 10 times the value of $R_S$. Anytime this condition is present, the voltage gain can be expected to be very close to, but never quite, unity.

The input impedance of the device is so high that, for practical applications, the circuit input impedance is the same as the value of $R_G$.

$$Z_i = R_G$$

The output resistance, $r_d$, of the common-drain transistor is approximately equal to $1/g_m$. This $r_d$ appears in parallel with the source resistance, $R_S$. So the output impedance for the circuit is

$$Z_o = \frac{r_d R_S}{r_d + R_S}$$

Using 3000 μmhos for $g_m$ and 10 kΩ for $R_S$ produces these results:

$$\begin{aligned}
r_d &= \frac{1}{g_m} \\
&= \frac{1}{3000 \times 10^{-6}} \\
&= 333 \ \Omega
\end{aligned}$$

$$Z_o = \frac{r_d R_S}{r_d + R_S}$$

$$= \frac{333 \times 10 \times 10^3}{333 + 10 \times 10^3}$$

$$= \frac{333 \times 10^4}{10{,}333}$$

$$= 322 \ \Omega$$

So when the value of $R_S$ is more than 10 times the value of $r_d$, the output impedance is approximately equal to $r_d$ or $1/g_m$.

## 3.4  JFET Biasing

The *biasing* arrangements for a JFET resemble electron-tube biasing circuits and have little association with biasing of bipolar transistors. The reason for this is the fact that both electron tubes and field-effect transistors are voltage-controlled devices. The JFET has no thermal-runaway problem. The two principal types of bias for the JFET are fixed bias and self-bias.

### 3.4.1  JFET Fixed Bias

The first step in designing any FET bias circuit is to determine the desired levels of $V_{DS}$ and $I_D$ under quiescent conditions. These are the conditions that exist when there is no input signal applied. The intersection of $V_{DS}$ and $I_D$ on the applicable characteristic graph reveals the level of $V_{GS}$ that will produce these conditions. Consider the circuit in Figure 3-24.

With the +24-V drain supply voltage indicated, maximum undistorted amplification can be obtained if the static $V_{DS}$ is +12 V, just half of the total supply voltage. So the desired static $V_{DS}$ is +12 V. The remaining 12 V must be dropped across $R_D$ by the static value of $I_D$. Since $R_D$ is 2.2 k$\Omega$, the static $I_D$ must be

$$I_D = \frac{12 \ \text{V}}{2.2 \ \text{k}\Omega}$$

$$= 5.5 \ \text{mA}$$

Now refer to Figure 3-25 and determine what value of $V_{GS}$ will produce an $I_D$ of 5.5 mA with 12 V of $V_{DS}$.

# Field-Effect Transistors

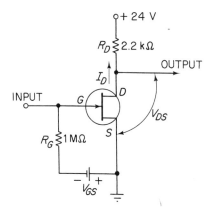

Figure 3-24.   JFET Fixed Bias

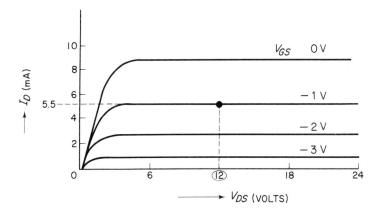

Figure 3-25.   $V_{GS}$ for $V_{DS} = 12$ V and $I_D = 5.5$ mA

The 12-V $V_{DS}$ line intersects the 5.5-mA $I_D$ line at a $V_{GS}$ of $-1$ V. This $-1$ V is the required bias for the circuit in Figure 3-24. The biasing circuit is completed by replacing the $V_{GS}$ battery with any steady voltage equal to 1 V. The polarity must be as indicated on the diagram.

### 3.4.2   JFET Self-Bias

The circuit of Figure 3-24 can be changed to a self-biasing circuit with only a few circuit modifications. The changes are shown in Figure 3-26. Notice that $R_G$ has been returned to ground and that a biasing

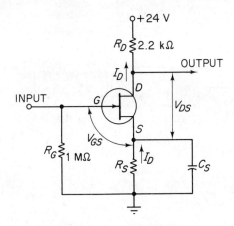

Figure 3-26. JFET Self-Bias

resistor, $R_S$, has been connected in series with the source. Capacitor $C_S$ across $R_S$ keeps a steady direct voltage on the source and bypasses the ac signal. This arrangement provides a steady value of bias, $V_{GS}$, but prevents signal degeneration.

Assuming the same desired conditions as in the previous example ($V_{DS} = 12$ V, $I_D = 5.5$ mA, and $V_{GS} = -1$ V), the only problem is to determine what value of $R_S$ will produce a voltage drop of 1 V.

$$R_S = \frac{V_{GS}}{I_D}$$

$$= \frac{-1 \text{ V}}{5.5 \text{ mA}}$$

$$= 180 \text{ } \Omega$$

When the 180-Ω resistor is inserted as $R_S$, $I_D$ becomes 5.5 mA and $V_{GS}$ becomes $-1$ V. There is still a 12-V drop across $R_D$, and this leaves 12 V from drain to ground. The $V_{DS}$, of course, is 12 V less the 1 V dropped across $R_S$. So there is a $V_{DS}$ of 11 V instead of the desired 12 V. This makes no substantial difference in the circuit operation. There is now a maximum swing of 11 V in both directions instead of 12 V.

## 3.5 MOSFET Biasing

Biasing for a MOSFET is exactly the same as for a JFET as long as fixed bias is concerned. However, pure self-bias for an enhancement MOSFET is impossible. If self-bias is used, it is combined with some form of fixed bias.

### 3.5.1 Enhancement-MOSFET Fixed Bias

In either a *P*- or an *N*-channel enhancement MOSFET, the polarity of the voltage, $V_{GS}$, must be such that it will attract the appropriate current carriers from the substrate and enhance the conduction channel between source and drain. With a *P*-channel MOSFET, $V_{GS}$ must be a negative potential in order to attract the positive current carriers. With an *N*-channel MOSFET, $V_{GS}$ must be a positive potential in order to attract the negative carriers. The circuit in Figure 3-27 is for fixed bias on an *N*-channel enhancement MOSFET. The procedure is the same as

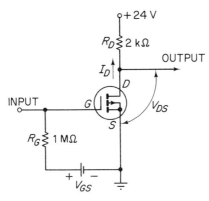

Figure 3-27. Enhancement-MOSFET Fixed Bias

previously described. First determine the desired quiescent values of $V_{DS}$ and $I_D$. Assume, in this case, that a maximum swing of the output is desired. This fixes the value of $V_{DS}$ as half the value of the supply voltage. With +24 V as a supply, $R_D$ must drop 12 V. With an $R_D$ of 2 kΩ, $I_D$ must be

$$I_D = \frac{V}{R_D}$$

$$= \frac{12 \text{ V}}{2 \text{ k}\Omega}$$

$$= 6 \text{ mA}$$

So the quiescent conditions are $V_{DS} = 12$ V and $I_D = 6$ mA. The graph in Figure 3-28 reveals the value of $V_{GS}$ that will produce these desired conditions. The intersection of $V_{DS} = 12$ V and $I_D = 6$ mA is on the +3-V $V_{GS}$ line. So +3 V of $V_{GS}$ will produce the desired conditions. The $V_{GS}$ battery is now replaced by any steady voltage that causes the gate to be 3 V more positive than the source.

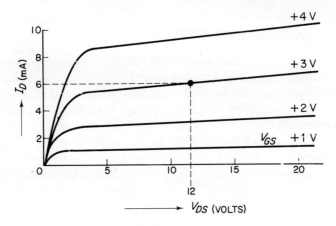

Figure 3-28.  $V_{GS}$ for $V_{DS} = 12$ V and $I_D = 6$mA

### 3.5.2  Enhancement-MOSFET Combination Bias

When self-bias is used with an enhancement MOSFET, it is combined with some form of fixed bias. One arrangement of this combination bias is illustrated in Figure 3-29. The fixed bias is provided by $R1$ in series with $R_G$. The self-bias is produced by $I_D$ through the source resistor, $R_S$. Current through $R_G$ from ground to $+V_{DD}$ produces a

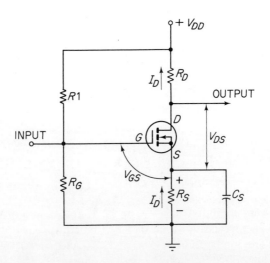

Figure 3-29.  Enhancement-MOSFET Combination Bias

# Field-Effect Transistors

positive potential between gate and ground. Current, $I_D$, from ground through $R_S$ produces a positive potential between source and ground. The voltage across $R_G$ tends to give $V_{GS}$ a positive potential, which it must have. The voltage across $R_S$ tends to give $V_{GS}$ a negative potential. So the effective $V_{GS}$ is the difference between the voltage across $R_G$ and the voltage across $R_S$. For proper operation this difference value must be equal to the predetermined $V_{GS}$, and the gate must be positive with respect to the source.

## 3.6 FET Symbols

|  |  |
|---|---|
| $A$ or $A_v$ | voltage amplification |
| $C_c$ | intrinsic channel capacitance |
| $C_{ds}$ | drain-to-source capacitance |
| $C_{gd}$ | gate-to-drain capacitance |
| $C_{gs}$ | gate-to-source capacitance |
| $C_{iss}$ | input capacitance, small signal, short circuit |
| $C_{oss}$ | output capacitance, small signal, short circuit |
| $C_{rss}$ | reverse transfer capacitance, small signal, short circuit |
| $e_n$ | equivalent input noise voltage |
| $g_c$ | forward conversion conductance |
| $g_{fs}$ | input inductance |
| $g_m$ | transconductance |
| $g_{os}$ | output inductance |
| $G$ or $G_{ps}$ | power gain |
| $G_{ps(c)}$ | conversion power gain |
| $I_D$ | direct drain current |
| $I_{DS(OFF)}$ | off current, drain to source |
| $I_{DSS}$ | zero-bias drain current |
| $I_{G1SS}$ | leakage current, gate 1 |
| $I_{G2SS}$ | leakage current, gate 2 |
| $I_{GSS}$ | gate leakage current |
| NF | spot noise |
| $r_d$ | active channel resistance |
| $r_{DS(ON)}$ | on resistance, drain to source |
| $R_{DS(OFF)}$ | off resistance, drain to source |
| $r_{gd}$ | leakage resistance, gate to drain |
| $r_{gs}$ | leakage resistance, gate to source |

| | |
|---|---|
| $r_i$ or $r_{iss}$ | input resistance |
| $r_o$ or $r_{oss}$ | output resistance |
| $r_s$ | effective gate series resistance |
| $V_{DB}$ | voltage, drain to substrate |
| $V_{DG1}$ | voltage, drain to gate 1 |
| $V_{DG2}$ | voltage, drain to gate 2 |
| $V_{DS}$ | voltage, drain to source |
| $V_{G1S}$ | voltage, gate 1 to source |
| $V_{G1S(OFF)}$ | cutoff voltage, gate 1 to source |
| $V_{G2S}$ | voltage, gate 2 to source |
| $V_{G2S(OFF)}$ | cutoff voltage, gate 2 to source |
| $V_{GB}$ | voltage, gate to substrate |
| $V_{GS}$ | voltage, gate to source |
| $V_{GS(OFF)}$ | cutoff voltage, gate to source |
| $V_o$ | offset voltage |

SECTION TWO

# DESIGN CONSIDERATIONS

This section deals with common factors faced in all types of solid-state circuit design. Many of the problems in design can be either eliminated or minimized by applying the information presented here. The information ranges from how to interpret data sheets, to transistor parameters, to thermal circuits, to essential test equipment, to spare parts.

CHAPTER FOUR

# Interpreting Data Sheets

The *data sheet* (also called *specification sheet*) that accompanies a solid-state device is fully as important to designers as the device itself. With proper interpretation, a designer obtains from this sheet nearly all the basic design information for that particular series of devices. In exacting work such as extremely high frequency, digital circuits, and very close tolerances, tests may still need to be performed under simulated circuit conditions. Most manufacturers of semiconductors publish data sheets of their products in either data books or catalogs. This readily available information enables a person to be thoroughly familiar with a product's characteristics before making a purchase. This chapter interprets a sample data sheet that was created for discussion purposes. After that, an actual data sheet from a leading manufacturer of semiconductor products is analyzed.

## 4.1 Sample Data Sheet

The data sheet in Figure 4-1 is a sample composed of imaginary data. It is intended for instruction only. Most data sheets will contain less information of the type shown, but they will have other information in the way of graphs and drawings. A transistor, or a series of transistors, is manufactured for specific applications. The intended applications are included on the data sheet. If a transistor should possess all the characteristics indicated in Figure 4-1, it could be used for high-speed switching, audio-frequency amplification, or radio-frequency amplification up to a frequency of about 200 MHz.

## TRANSISTOR SPECIFICATIONS

### Maximum Ratings (25° C)

Voltages (dc)
    Collector to base                     $V_{CB}$            40 V
    Emitter to base                       $V_{EB}$            5 V
    Collector to emitter              $V_{CEO}$          40 V

Collector Current (dc)               $I_C$            200 mA

Power (Above 25°C derate 9.1 mW/°C)
    Collector dissipation              $P_C$           1.6 W

Temperature
    Junction                                 $T_J$           + 200°C
    Storage range                      $T_{STG}$          − 65 to + 200°C

### Electrical Characteristics (25°C)

| | | MIN | MAX |
|---|---|---|---|
| Small Signal | | | |
|   Forward current transfer ratio | $h_{fe}$ | 2.5 | |
|   Input impedance | $h_{ie}$ | 500 ohms | 1500 ohms |
|   Reverse voltage transfer ratio | $h_{re}$ | | $5 \times 10^4$ |
|   Output admittance | $h_{oe}$ | | 1.2 µmhos |
|   Noise figure | NF | | 28 dB |
|   Current gain · BW product | $f_T$ | 250 | |
|     ($I_C = -20$ mA, $V_{CE} = -20$ V) | | | |
| High frequency | | | |
|   Frequency cutoff | | | |
|     ($V_{CB} = -5$ V, $I_E = -1$ mA) | $f_{ab}$ | | 15 MHz |
|   Collector to base capacitance | | | |
|     ($V_{CB} = -10$ V, $I_E = 0$, $f = 100$ kHz) | $C_{ob}$ | | 4 pF |
| Direct Current | | MIN | MAX |
|   Collector-base breakdown voltage | $BV_{CBO}$ | 40 V | |
|     ($I_C = -10$ µA, $I_E = 0$) | | | |
|   Collector cutoff current | | | |
|     ($V_{CB} = -20$ V, $I_E = 0$) | $I_{CBO}$ | | 0.025 µA |
|     ($V_{CB} = -20$ V, $I_E = 0$, $T_A = 150°C$) | | | 30 µA |
|   Collector-emitter saturation voltage | | | |
|     ($I_C = -50$ mA, $I_B = -5$ mA) | $V_{CE\,(sat)}$ | | 0.5 V |
|   Forward current transfer ratio | | | |
|     ($I_C = -10$ mA, $V_{CE} = -10$ V) | $h_{FE}$ | 25 | |
|   Collector-emitter breakdown voltage | | | |
|     ($I_C = -10$ mA, $I_B = 0$) | $BC_{CEO}$* | 20 V | |
| Switching ($I_{B1} = 0.4$ mA, $I_{B2} = -0.4$ mA, $I_C = 2.8$ mA) | | | |
|   Delay time | $t_d$ | | 0.75 µS |
|   Rise time | $t_r$ | | 0.5 µS |
|   Storage time | $t_s$ | | 0.05 µS |
|   Fall time | $t_f$ | | 0.15 µS |

*Pulse test: PW $\leq 300$ µS, Duty cycle $\leq 2\%$

Figure 4-1. Sample Data Sheet

### 4.1.1 Data Groups

There are two important groups of data on the sheet: maximum ratings and electrical characteristics. Notice that both of these groups specify a temperature of 25°C. This is room temperature; other temperature conditions will alter the performance of the device. It is important to remember that, unless otherwise specified, all ratings and characteristics are understood to be the performance standards at 25°C.

The *maximum ratings* section includes voltages, current, power, and temperature. Some data sheets specify absolute maximum ratings. These given values must be considered as absolute maximums, whether or not it is so specified. Exceeding one or more of these maximum ratings not only causes inferior performance but also seriously jeopardizes the transistor.

The *electrical characteristics* are subdivided into small-signal characteristics, high-frequency characteristics, direct-current characteristics, and switching characteristics. Some of these headings are generally omitted, depending upon the intended application of the device. For example, when a device is not recommended for switching, there would be little point in publishing switching characteristics.

### 4.1.2 Maximum Voltages

Maximum voltages are dc values, and they may be given in terms of operating voltage or test voltage. In this case, the collector-to-base and the emitter-to-base voltages are operating voltages. The collector-to-emitter voltage is a test voltage. The test voltage is revealed by the O in the symbol. The letters $V_{CEO}$ mean collector-to-emitter voltage with the base open. It must be remembered that a cutoff or near-cutoff transistor has the full value of the supply voltage, $V_{CC}$, on the collector. If the collector load impedance is a coil, the value of $V_C$ may actually exceed the value of $V_{CC}$. These are important items for consideration in order to prevent exceeding the maximum voltage limits.

All power supplies, with the exception of batteries, tend to fluctuate under changing load conditions. The voltage variations must be planned for. The emitter-to-base voltage is listed as a direct voltage, but the signal voltage must also be considered. The 5 V is an operating voltage and should not be exceeded during any part of the input signal.

### 4.1.3 Collector Current

Current, temperature, and power are all interrelated. As current increases, power increases and temperature increases. An increase in temperature causes an increase in current and an increase in power. For any temperature in excess of 25°C, the maximum collector-current rating should include a wide margin of safety.

In many cases, a transistor can be destroyed by excessive heat, with both voltage and current well within the maximum limits. For example, if this transistor has 200 mA of collector current and 20 V of collector voltage, it would have to dissipate a power of 4 watts (W), whereas its maximum power rating is 1.6 W. In this example, the transistor's ambient temperature was still 25°C. At higher temperatures the dangers increase. The best policy in design is to make sure that no combination of $I_C$ and $V_C$ can cause the power to exceed the maximum rating.

### 4.1.4 Power

Notice that the power rating contains a derating factor. The transistor can dissipate 1.6 W of power at 25°C. Its power limit decreases by 9.1 mW for each °C over 25°. The maximum operating junction temperature is +200°C. However, at 200°C the power-handling ability of this transistor is reduced from 1.6 W to 7.5 mW.

### 4.1.5 Temperature

It would be naive to expect a transistor to operate for any considerable time at the ideal temperature of 25°C. This simply is *not* a normal environment. The effects of higher temperatures must be kept in mind. In circuits where temperature can possibly become a problem, heat sinks should be used to aid in dissipating the excessive heat.

The transistor represented by Figure 4-1 can be operated at any temperature up to 200°C, but the power rating must be adjusted accordingly.

### 4.1.6 Small-Signal Characteristics

The small-signal characteristics of a transistor are a starting point for circuit design, but they fall short of a truly sound basis for constructing practical circuits. These small-signal parameters are obtained under a fixed set of operating conditions. When the constructed circuit fails to exactly duplicate these conditions, the parameters change.

The parameters listed here are straightforward hybrid parameters for a common-emitter configuration. The values indicated were obtained at predetermined values of collector-to-base voltage and emitter current. Different manufacturers call the same parameters by different names and sometimes designate them by different symbols.

The noise figure and current gain–bandwidth product become important in most amplifier applications.

### 4.1.7 High-Frequency Characteristics

High-frequency characteristics are likely to appear only when a transistor is a suitable device for amplifying frequencies in the radio-frequency (RF) band. However, in the broad sense, the RF band is almost any frequency above about 20 kHz.

The figures listed here are very inflexible. The voltage, current, and test frequencies are specified. The RF circuit designer must know the frequency response over a band of frequencies. In addition to the figures shown here, most manufacturers provide graphs that show response over a specified frequency range.

### 4.1.8 Direct-Current Characteristics

Notice that most dc characteristics are test values, and each one has a rigid set of conditions. This gives designers a starting point in the design of bias circuits, but these values will change drastically with a change in voltage, current, or temperature. Generally, dc characteristics are not critical to the final circuit operation.

### 4.1.9 Switching Characteristics

Switching characteristics are critical factors in the design of all types of pulse circuits. The time factors specified here determine the operating limits of switching circuits. All these factors are characteristics of the output pulse in relation to the input pulse. The exact relationships are illustrated in Figure 4-2.

*Time delay* is the time from the leading edge of the input pulse until the output has risen to 10 percent of its maximum amplitude.

*Rise time* is the time required for the output to rise from 10 percent of its maximum amplitude to 90 percent of this maximum amplitude.

*Storage time* is the time from the trailing edge of the input until the output drops to 90 percent of its maximum value.

*Fall time* is the time required for the output to drop from 90 percent of its maximum amplitude to 10 percent of this maximum value.

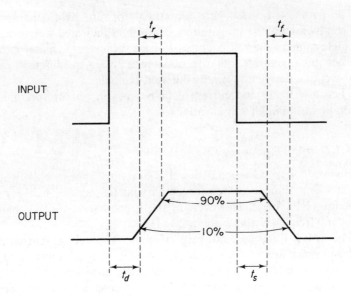

Figure 4-2.  Switching Characteristics

## 4.2   Actual Sample Data Sheet

The data sheet in Figure 4-3 is an actual specification sheet for General Electric's D41D transistor. This data sheet and the remainder of this chapter were originally published in General Electric's *Transistor Manual*. The information is reprinted here by special permission of General Electric Company, Semiconductor Products Department, Auburn, New York. Only the figure numbers have been changed.

### 4.2.1   Part 1—The Transistor Specification Sheet

The published transistor specification sheet is fully as important as the device it describes since it provides the description necessary for sensible use of the subject transistor.

Four general categories of information are presented. These are:

1. a statement of broad device capabilities and intended service
2. absolute maximum ratings
3. electrical characteristics
4. generic or family electrical characteristics

Let's study each of these categories in some detail and use the specifications sheet for General Electric D41D (Figure 4-3) as a guide.

1. GENERAL DEVICE CAPABILITIES

The lead paragraph found at the top of the sheet furnishes the user with a concise statement of the most likely applications and salient electrical characteristics of the device. It is useful in first comparison of devices as one selects the proper device for a particular application.

2. ABSOLUTE MAXIMUM RATINGS

Absolute maximum ratings specify those electrical, mechanical, and thermal ratings of a semiconductor device which, as limiting values, define the maximum stresses beyond which either initial performance or service life is impaired.

*Voltage*

The voltages specified in the *Absolute Maximum Ratings* portion of the sheet are breakdown voltages with reverse voltage applied to one selected junction, or across two junctions with one junction reverse biased and the second junction in some specified state of bias. Single junction breakdown either between collector and base or between emitter and base has the form shown in Figure 4-4.

The solid portion of the curve is the active, normally used portion of a diode or any compound junction device. The dotted portion exhibits large dramatic changes in reverse current for small changes in applied voltage. This region of abrupt change is called the *breakdown* region. If breakdown occurs at relatively low voltage, the mechanism is through tunneling or "zener" breakdown. The means of conduction is through electrons which have "tunneled" from valence to conduction energy levels.

At higher voltage levels conduction is initiated and supported by solid ionization. When the junction is reverse biased, minority current (leakage current) is made up of holes from the $n$-type material and electrons from the $p$-type material. The high field gradient supplies carriers with sufficient energy to dislodge other valence electrons, raising their energy level to the conduction band resulting in a chain generation of hole-electron pairs. This process is called *avalanche*. While theory predicts an abrupt, sharp (sometimes called *hard*) characteristic in the breakdown region, a *soft* or gradual breakdown often occurs. Another possibility is the existence of a negative resistance "hook."

# Silicon Power Tab Transistors
## For Consumer and Industrial Electronics
### "COLOR MOLDED"

**PNP D41D** (Previously D31B)

50.64 6/69
Supersedes 50.64 3/69

Leads Can Be Formed To A TO-5 Pin Configuration

The D41D is a black, silicone plastic and encapsulated, power transistor designed for output stages of medium power stereo, phonos and other audio amplifiers and as drivers for very high power amplifiers. There are also many general purpose industrial and consumer applications.

**FEATURING:**
- PNP complement to D40D NPN (previously D28D)
- Black for PNP, brown for NPN
- Low collector saturation voltage ($-0.5V$ typ. @ $-1.0A$ $I_C$)
- Excellent linearity
- Fast switching

## absolute maximum ratings: (25°C) (Unless Otherwise Specified)

| Voltages | | D41D1(1) D41D2 | D41D4 D41D5 | D41D7 D41D8 | Units |
|---|---|---|---|---|---|
| Collector to Emitter | $V_{CEO}$ | $-30$ | $-45$ | $-60$ | Volts |
| Emitter to Base | $V_{EBO}$ | $-5$ | $-5$ | $-5$ | Volts |
| Collector to Emitter | $V_{CES}$ | $-45$ | $-60$ | $-75$ | Volts |

88

| | | | | | |
|---|---|---|---|---|---|
| **Current**(2) | | | | | |
| Collector (Continuous) | $I_C$ | −1 | −1 | −1 | Amps |
| Collector (Peak) | | −1.5 | −1.5 | −1.5 | Amps |
| **Power Dissipation**(2) | | | | | |
| Tab at 25°C | $P_T$ | 6 | 6 | 6 | Watts |
| Tab at 70°C | | 4 | 4 | 4 | Watts |
| Free Air at 25°C | | | | | |
| With Tab | | 1.25 | 1.25 | 1.25 | Watts |
| Without Tab | | 1 | 1 | 1 | Watts |
| Free Air at 50°C | | | | | |
| With Tab | | 1 | 1 | 1 | Watts |
| Without Tab | | 0.75 | 0.75 | 0.75 | Watts |

## electrical characteristics: (25°C) (Unless Otherwise Specified)

| | | D41D1 D41D4 D41D7 | | D41D2 | | D41D5 D41D8 | | |
|---|---|---|---|---|---|---|---|---|
| **Forward Current Transfer Ratio** | | Min. | Max. | Min. | Max. | Min. | Max. | |
| ($V_{CE}$ = −2V, $I_C$ = −100 mA) | $h_{FE}$ | 50 | 150 | 120 | 300 | 120 | 360 | |
| ($V_{CE}$ = −2V, $I_C$ = −1A) | $h_{FE}$ | 10 | — | 20 | — | 10 | - | |
| **Collector to Emitter Voltage** | | | | | | | | |
| ($I_C$ = −10 mA) D41D1, 2 | $V_{CEO}$ | | −30 | — | | — | | Volt |
| D41D4, 5 | $V_{CEO}$ | | −45 | — | | — | | Volt |
| D41D7, 8 | $V_{CEO}$ | | −60 | — | | — | | Volt |

INTEGRATED CIRCUITS PROJECT
ELECTRONICS PARK, SYRACUSE, N.Y. 13201

GENERAL ELECTRIC

In Canada, Canadian General Electric Company Ltd., Toronto, Ont.
Outside the U.S.A. and Canada, by: Electronic Component Sales,
I.G.E. Export Division, 159 Madison Avenue, New York, N.Y. 10016

Figure 4-3. Actual Data Sheet

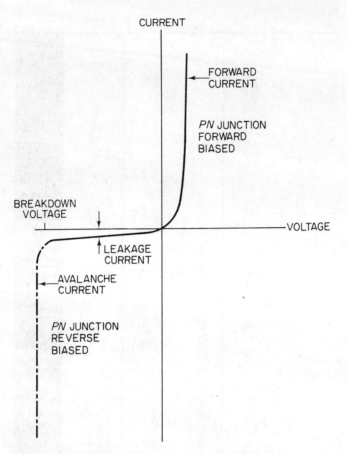

Figure 4-4. Single Junction Breakdown

The hook usually occurs when zener breakdown is the predominant mechanism. Figure 4-5 graphically illustrates these possibilities. In practice, silicon, because of lower leakage current, exhibits a sharper knee than does germanium.

The family of the D40D and D41D silicon devices are measured for individual junction breakdown voltages at a current of 0.1 microampere; $V_{CBO}$, the collector-base diode breakdown voltage—with emitter open circuited or floating—is a minimum of 75 volts for the D41D7 and 8.

$V_{EBO}$, the emitter-base breakdown voltage—with collector open circuited or floating—is specified at 5 volts minimum for the D41D.

Interpreting Data Sheets

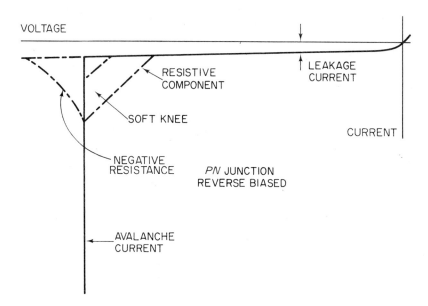

Figure 4-5. Typical Variations in Breakdown Characteristic of a PN Junction

The breakdown voltage between collector and emitter is a more complex process. The collector-base junction in any configuration involving breakdown is always reverse biased. On the other hand, the condition applied to the emitter-base diode depends upon the nature of base lead connection. The most stringent requirement is realized by allowing the base to float. The next most stringent requirement is connecting the base to the emitter through a resistor. A more lenient measurement is with base and emitter shorted. Finally, the condition yielding the highest breakdown voltage is that which applies reverse bias to the emitter-base junction. The symbols for the breakdown voltage, collector to emitter, under the foregoing base connection conditions are $V_{CEO}$, $V_{CER}$, $V_{CES}$, and $V_{CEX}$ respectively. On some specification sheets the letter $B$, signifying breakdown, precedes the voltage designation, i.e., $BV_{CEO}$.

The generic shape of breakdown characteristics differs among transistors fabricated by different processes. Figure 4-6 is typical of the planar epitaxial process. Note that the $BV_{CEO}$ curve exhibits little current flow ($I_{CEO}$) until breakdown is initiated. At breakdown a region of negative resistance appears and disappears at increased collector voltage. The region of negative resistance is not suitable for measurement and specification because of instability. The low current positive resistance re-

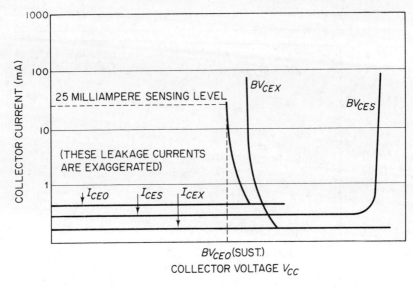

Figure 4-6.  Typical Planar Epitaxial Collector Breakdown Characteristics

gion below (in voltage) the breakdown region is so low as to cause instrumentation difficulties. It is desirable, therefore, to measure breakdown voltage at a current, in the breakdown region, where the slope is positive. This current for the D41D family is 10 mA.

Since $BV_{CER}$ and $BV_{CEX}$ as well as $BV_{CEO}$ exhibit a negative resistance region, they must also be measured in a region of positive resistance. The voltage thus measured is always less than voltage needed to establish breakdown. For this reason it has been suggested that these voltages be named differently than breakdown voltages. One proposal is to designate them as "sustaining" voltages with the prefix letter $L$ substituted for $B$, i.e., $LV_{CER}$. The nomenclature $V_{CER(SUST.)}$ has also been used.

The behavior of alloy devices is sufficiently different to warrant separate consideration. Figure 4-7 illustrates a typical family of breakdown characteristics. Since leakage currents are appreciable in this class of devices, they form an important part of breakdown consideration. In the specification of $BV_{CEO}$, consideration must be given to $I_{CO}$ multiplication. In this connection $I_{CBO}$ is approximately $h_{FE} \times I_{CO}$. This product may exceed 100 $\mu$A (the usual $BV_{CBO}$ sensing current) at voltages well below breakdown. For this reason it is common to specify breakdown at a collector current of 600 $\mu$A. Figure 4-7 shows the realistic

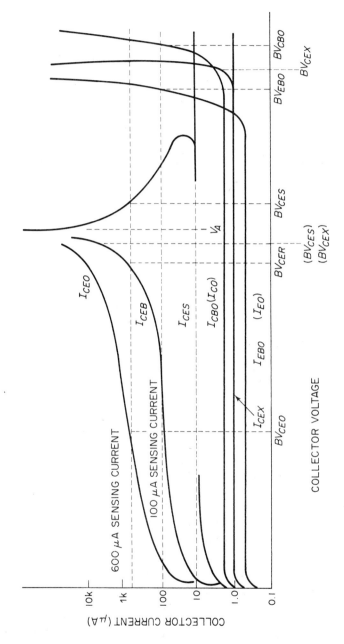

Figure 4-7. Typical Family of Alloy Transistor Breakdown Characteristics

increase in voltage resulting from the use of a 600 $\mu$A sensing current. The earlier statement that $BV_{CEO}$ is a very conservative rating is particularly true of germanium alloy devices. It is primarily applicable to circuits with little or no stabilization.

$BV_{CES}$ is measured with the base shorted to the emitter. It is an attempt to indicate more accurately the voltage range in which the transistor is useful. In practice, using a properly stabilized circuit, the emitter junction is normally forward biased to give the required base current. As temperature is increased, the resulting increase in $I_{CO}$ and $h_{FE}$ requires that the base current decrease if a constant, i.e., stabilized, emitter current is to be maintained. In order that base current decrease, the forward bias voltage must decrease. A properly designed biasing circuit performs this function. If temperature continues to increase, the biasing circuit will have to reverse bias the emitter junction to control the emitter current. $V_{BE} = 0$ is identically the same condition as a base to emitter short as far as analysis is concerned. Therefore, the $BV_{CES}$ rating indicates what voltage can be applied to the transistor when the base and emitter voltages are equal, regardless of the circuit or environmental conditions responsible for making them equal. There is a negative resistance region associated with $I_{CES}$. At sufficiently high currents the negative resistance disappears. Oscillations may occur depending on the circuit stray capacitance and the circuit load line. In fact "avalanche" transistor oscillators are operated in just this mode.

Conventional circuit designs must avoid these oscillations. If the collector voltage does not exceed $V_A$ (Figure 4-7) there is no danger of oscillation. $V_A$ is the voltage at which the negative resistance disappears at high current.

To avoid the problems of negative resistance associated with $BV_{CES}$, $BV_{CER}$ was introduced. The base is connected to the emitter through a specified resistor. This condition falls between $BV_{CEO}$ and $BV_{CES}$ and for most germanium alloy transistors avoids creating a negative resistance region. For most low power transistors the resistor is 10,000 ohms. The significance of $BV_{CER}$ requires careful interpretation. At low voltages the resistor tends to minimize the collector current. Near breakdown the resistor becomes less effective permitting the collector current to increase rapidly.

Both the value of the base resistor and the voltage to which it is returned are important. If the resistor is connected to a forward biasing voltage the resulting base drive may saturate the transistor giving the illusion of a collector to emitter short. Returning the base resistor to the emitter voltage is the standard $BV_{CER}$ test condition. If the resistor is returned to a voltage which reverse biases the emitter junction, the collector current will approach $I_{CO}$. For example, many computer circuits use an emitter reverse bias of about 0.5 volt to keep the collector current at cut-off. The available power supplies and desired circuit functions determine the value of base resistance. It may range from 100 to 100,000 ohms with equally satisfactory performance provided the reverse bias voltage is maintained.

In discussing the collector to emitter breakdown so far, in each case the collector current is $I_{CO}$ multiplied by a circuit dependent term. In other words all these collector to emitter breakdowns are related to the collector junction breakdown. They all depend on avalanche current multiplication.

Another phenomenon associated with collector to emitter breakdown is that of *reach-through or punch-through*. Silicon devices as typified by grown diffused, double diffused, planar, mesa, and planar epitaxial structures do not exhibit this characteristic. The phenomenon of reach-through is most prevalent in alloy devices having thin base regions, and lighter base region doping than collector region doping. As reverse voltage is increased, the depletion layer spreads more in the base than in the collector and eventually "reaches" into the emitter. An abrupt increase in current results.

The dotted lines in Figure 4-7 indicate the breakdown characteristics of a reach-through limited transistor. Several methods are used to detect reach-through. $BV_{CEX}$ (breakdown voltage collector to emitter with base reverse biased) is one practical method. The base is reverse biased by 1 volt. The collector current $I_{CEX}$ is monitored. If the transistor is avalanche limited, $BV_{CEX}$ will approach $BV_{CBO}$. If it is reach-through limited it will approach $BV_{CES}$.

Note that $I_{CEX}$ before breakdown is less than $I_{CO}$. Therefore, if $I_{CO}$ is measured at a specified test voltage and then the emitter is connected with a reverse bias of 1 volt, the $I_{CO}$ reading will decrease if reach-through is above the test voltage and will increase if it is below.

"Emitter floating potential" is another test for reach-through. If the voltage on an open-circuited emitter is monitored while the collector to base voltage is increased, it will remain within 500 mV of the base voltage until the reach-through voltage is reached. The emitter voltage then increases at the same rate as the collector voltage. $V_{RT}$ is defined as $(V_{CS} - 1)$ where $V_{CS} = 1$ V.

*Current*

The absolute maximum collector current, shown as 1.5 amperes for the D41D, is a pulse current rating. In this case it is the maximum collector current for which $h_{FE}$ is specified. In some cases the current level at which $h_{FE}$ drops from its maximum value by 50% is specified. In all cases judgement concerning adverse life effects is a major consideration.

Also in all cases no other absolute maximum rating can be exceeded in using this rating. In cases of very short, high current pulses, the power dissipated in transition from cutoff to saturation must be considered so that thermal ratings are not exceeded.

### Transistor Dissipation

Transistor dissipation ratings are thermal ratings, verified by life test, intended to limit junction temperature to a safe value. Device dissipation is shown for case temperature (heat sink usage) and free air temperature without a heat sink. The specification sheet gives the derating in transistor dissipation for other temperatures. This thermal derating factor can be interpreted as the absolute maximum thermal conductance junction to air or case, under the specified conditions. If dissipation is specified at 25°C case temperature, an infinite heat sink is implied and the dissipation is at its largest allowable value.

Both free air and heat sink ratings are valuable since they give limit application conditions from which intermediate (in thermal conductance) methods of heat sinking may be estimated.

### Temperature

The D41D family carries a storage temperature rating extending from −55°C to +150°C. High temperature storage life tests substantiate continued compliance with the upper temperature extreme. Further, the mechanical design is such that thermal/mechanical stresses generated by rated temperature extremes cause no electrical characteristic degradation.

Operating junction temperature although stated implicitly by thermal ratings may also be stated explicitly as an absolute maximum junction temperature.

### 3. ELECTRICAL CHARACTERISTICS

Electrical characteristics are the important properties of a transistor which are controlled to insure circuit interchangeability and describe electrical parameters.

### DC Characteristics

Forward current transfer ratio, $h_{FE}$, is specified over one decade of collector current from 100 milliamperes to 1 ampere.

The $h_{FE}$ measurements are made at a 2% duty cycle and pulse widths less than or equal to 300 microseconds. This precaution is necessary to prevent an increase in junction temperature as the current is increased.

The second characteristic shown is the voltage ratings, repeated in the order of the absolute maximum ratings, but this time showing the condition of test. Note that the electrical parameters are measured at 25°C ambient temperature unless otherwise noted.

The characteristics below are shown on the second page of the D41D specification. Base saturation, $V_{BE(SAT)}$, specifies the base input voltage characteristic under the condition of both junctions being forward biased. The conditions of measurement specify a base current and a collector current. This parameter is of particular interest in switch designs.

Collector saturation voltage, $V_{CE(SAT)}$, is the electrical characteristic describing the voltage drop from collector to emitter with both base-emitter and collector-base junctions forward biased. Base and collector currents are stipulated. The quotient of collector and base currents is termed "forced beta."

A wide difference in $V_{BE(SAT)}$ and $V_{CE(SAT)}$ is undesirable if Darlington connection of devices is desired for saturated switching. The collector saturation characteristic of the compound device demonstrates that the lead section is incapable of saturating the output section. Modification of the circuit to provide separate connection of the input section collector directly to the joint collector supply will provide the needed $V_{BE(SAT)}$ to allow output section saturation.

### *Cutoff Characteristics*

This primarily concerns transistor leakage currents. This examination deals with phenomena which predominate in alloy structures. The principal differences in planar epitaxial devices lie in the relative magnitudes of the leakage current components. The complete protection afforded by the passivation layer reduces surface leakage to a very small value. Further, it reduces the surface thermal component by decreasing recombination velocity. At high temperatures planar devices follow predicted behavior quite well. At lower temperatures, the temperature rate is considerably less than that which would be predicted by the theoretical model.

### *High Frequency Characteristics*

The small-signal forward current transfer ratio, $h_{fe}$, or gain-band width product, $f_T$, is specified for high frequency amplifier applications.

### *Switching Characteristics*

Switching characteristics are transient response times $t_d$, $t_r$, $t_s$, and $t_f$. The specified maximum rise, storage, and fall times are measured in a specified circuit. The switching times measured are highly circuit dependent.

The transistor specification sheet is, without doubt, the most important work tool the electronics circuit designer has at his disposal when it is understood and used intelligently.

CHAPTER FIVE

# Common Problems in Design

One of the first problems in solid-state circuit design is choosing a device that will do the required job. When choosing a transistor it is important to be able to predict how that transistor will react in a particular circuit. One approach is use of load lines on characteristics graphs; another is use of specified parameters and equivalent circuits. The load-line approach works very well when dealing with large signals. The parameter–equivalent circuit method is more reliable for small signals. A small signal is a signal that is small in comparison to the dc bias. A large signal is large by the same comparison. This section deals primarily with parameters and equivalent circuits, but it also discusses other types of problems normally encountered in circuit design.

## 5.1 Basic Parameters

A *parameter* is a quantity whose value varies with the circumstances of its application. A transistor has four variable quantities: input current, output current, input voltage, and output voltage. There are three types of parameters, and they are measured by placing the transistor in a specified test configuration. The three types are: open circuit, short circuit, and hybrid. The hybrid parameters are obtained by a combination of open-circuit and short-circuit conditions.

### 5.1.1 Open-Circuit Parameters

*Open-circuit parameters* are resistance quantities and are obtained by using the two currents as independent variables. The independent variables are the ones that are opened. The prime symbol for open-circuit parameters is $r$ for resistance. Sometimes impedance is used instead of resistance and $z$ is substituted for $r$. The four open-circuit parameters are:

$r_i$ = input resistance with output open

$r_r$ = reverse transfer resistance with input open

$r_f$ = forward transfer resistance with output open

$r_o$ = output resistance with input open

Another subscript letter is added to each parameter to designate the circuit configuration. When this third letter is $b$, the circuit is a common base. An $e$ indicates a common emitter, and $c$ designates common collector. Here are the four open-circuit parameters for each configuration:

| CB | CE | CC |
|---|---|---|
| $r_{ib}$ | $r_{ie}$ | $r_{ic}$ |
| $r_{rb}$ | $r_{re}$ | $r_{rc}$ |
| $r_{fb}$ | $r_{fe}$ | $r_{fc}$ |
| $r_{ob}$ | $r_{oe}$ | $r_{oc}$ |

Parameters for any one configuration are easily converted to either of the other configurations. This conversion process is covered later in this chapter. Meanwhile, for the sake of simplicity, the discussion concerns parameters for *common-emitter* configurations.

Each parameter is obtained by placing the transistor in a specified test circuit and taking measurements. A test circuit for each of the four open-circuit parameters is illustrated in Figure 5-1. The input resistance with the output open is $r_{ie}$. Test circuit a is used for this measurement. The ratio of a change in $V_{be}$ to a change in $I_b$ is $r_{ie}$:

$$r_{ie} = \frac{\Delta V_{be}}{\Delta I_b}$$

The reverse transfer resistance with an open input is $r_{re}$. The measurements are taken from test circuit b. The ratio of a change in $V_{be}$ to a change in $I_c$ is $r_{re}$:

$$r_{re} = \frac{\Delta V_{be}}{\Delta I_c}$$

Common Problems in Design

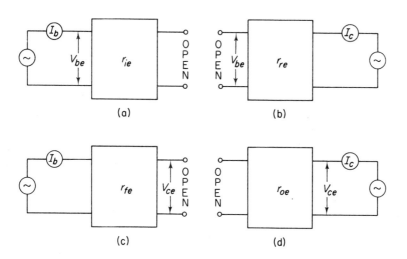

Figure 5-1.  Test Circuits for Open-Circuit Parameters

The forward transfer resistance with an open output is $r_{fe}$. Test circuit c is used for these measurements. The ratio of a change in $V_{ce}$ to a change in $I_c$ is $r_{fe}$:

$$r_{fe} = \frac{\Delta V_{ce}}{\Delta I_c}$$

The output resistance when the input is open is $r_{oc}$. Test circuit d is used for these measurements. The ratio of a change in $V_{ce}$ to a change in $I_c$ is $r_{oe}$:

$$r_{oe} = \frac{\Delta V_{ce}}{\Delta I_c}$$

Keep in mind that these are dynamic ratios. These circuits are open as indicated to ac, but the dc circuit is complete in each case. The voltage and current values in the equations are alternating values. Notice in each equation that voltage is divided by current. This is a simple application of Ohm's law for calculating resistance:

$$R = \frac{E}{I}$$

### 5.1.2 Short-Circuit Parameters

*Short-circuit parameters* are created by making the two voltages independent variables. This is done by connecting ac shorts across specified portions of the test circuit. Just as shorts and opens are opposite conditions, open-circuit parameters and short-circuit parameters are reciprocals of one another. The short-circuit parameters are in terms of conductance. The symbol is $g$ and the unit of measure is the mho. Sometimes admittance is used instead of conductance. When admittance is used, $y$ is substituted for $g$. Here are the four short-circuit parameters:

$g_i$ = input conductance with output shorted
$g_r$ = reverse transfer conductance with input shorted
$g_f$ = forward transfer conductance with output shorted
$g_o$ = output conductance with input shorted

Again, a third letter is used to specify the circuit configuration. Here are the four short-circuit parameters for each configuration:

| CB | CE | CC |
|---|---|---|
| $g_{ib}$ | $g_{ie}$ | $g_{ic}$ |
| $g_{rb}$ | $g_{re}$ | $g_{rc}$ |
| $g_{fb}$ | $g_{fe}$ | $g_{fc}$ |
| $g_{ob}$ | $g_{ob}$ | $g_{oc}$ |

The test circuits for obtaining short-circuit parameters are illustrated in Figure 5-2.

The input conductance with the output shorted is $g_{ie}$. This information is obtained with test circuit a. The ratio of a change in $I_b$ to a change in $V_{be}$ is $g_{ie}$:

$$g_{ie} = \frac{\Delta I_b}{\Delta V_{be}}$$

The reverse transfer conductance with the input shorted is $g_{re}$. The values for determining $g_{re}$ are taken from test circuit b. The ratio of a change in $I_b$ to a change in $V_{ce}$ is $g_{re}$:

$$g_{re} = \frac{\Delta I_b}{\Delta V_{ce}}$$

# Common Problems in Design

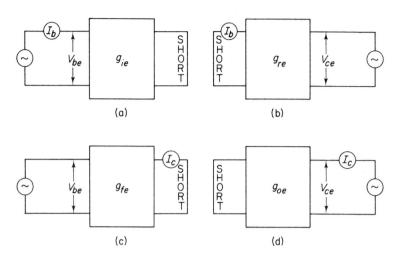

Figure 5-2. Test Circuits for Short-Circuit Parameters

The forward transfer conductance with the output shorted is $g_{fe}$. Test circuit c provides these values. The ratio of a change in $I_c$ to a change in $V_{be}$ is $g_{fe}$:

$$g_{fe} = \frac{\Delta I_c}{\Delta V_{be}}$$

The output conductance with a shorted input is $g_{oe}$. These values are obtained from test circuit d. The ratio of a change in $I_c$ to a change in $V_{ce}$ is $g_{oe}$:

$$g_{oe} = \frac{\Delta I_c}{\Delta V_{ce}}$$

Notice that each short-circuit parameter is a reciprocal of the corresponding open-circuit parameter. In each case, the conductance is

$$G = \frac{I}{E}$$

### 5.1.3 Hybrid Parameters

The *hybrid parameters* are mixed parameters. They are created by choosing one voltage and one current as independent variables. Hybrid parameters can be developed from voltage–current characteristics graphs, but they are generally specified on the transistor data sheet.

The data sheet lists the parameters for only one of the three basic circuit configurations, but it is easy to convert to either of the other two. The four hybrid parameters are:

$$h_i = \text{input resistance}$$
$$h_r = \text{reverse voltage gain}$$
$$h_f = \text{forward current gain}$$
$$h_o = \text{output conductance}$$

The resistance parameter, $h_i$, is measured in ohms, and the conduction, $h_o$, is in mhos. The remaining two, $h_r$ and $h_f$, are ratios, so they have no units.

A third letter is added, as before, to specify the circuit configuration. Here are the four hybrid parameters for each configuration:

| CB | CE | CC |
| --- | --- | --- |
| $h_{ib}$ | $h_{ie}$ | $h_{ic}$ |
| $h_{rb}$ | $h_{re}$ | $h_{rc}$ |
| $h_{fb}$ | $h_{fe}$ | $h_{fc}$ |
| $h_{ob}$ | $h_{oe}$ | $h_{oc}$ |

A hybrid parameter is a voltage–current ratio under specified conditions. Input resistance is $h_{ie}$. Input resistance is a ratio of a change in $V_{be}$ to a change in $I_b$ at a constant value of $V_{ce}$; this is input voltage divided by input current:

$$h_{ie} = \frac{\Delta V_{be}}{\Delta I_b} \qquad (V_{ce} \text{ constant})$$

Reverse voltage gain is $h_{re}$. Reverse voltage gain is a ratio of a change in $V_{be}$ to a change in $V_{ce}$ at a constant value of $I_b$. This is input voltage divided by output voltage:

$$h_{re} = \frac{\Delta V_{be}}{\Delta V_{ce}} \qquad (I_b \text{ constant})$$

Forward current gain is $h_{fe}$. Forward current gain is a ratio of a change in $I_c$ to a change in $I_b$ at a constant value of $V_{ce}$. This is output current divided by input current:

$$h_{fe} = \frac{\Delta I_c}{\Delta I_b} \qquad (V_{ce} \text{ constant})$$

# Common Problems in Design

Current gain for a common-base circuit is alpha; for a common emitter, it is beta. Therefore,

$$h_{fb} = \text{alpha}$$

$$h_{fe} = \text{beta}$$

Output conductance is $h_{oe}$. Output conductance is a ratio of a change in $I_c$ to a change in $V_{ce}$ at a constant value of $I_b$. This is output current divided by output voltage:

$$h_{oe} = \frac{\Delta I_c}{\Delta V_{ce}} \quad (I_b \text{ constant})$$

The hybrid-parameter values vary over a wide range as different transistors are considered. For this reason, there is no typical set of parameter values. With a given transistor, there is a wide difference in parameters when all three circuit configurations are considered. Here is a sample set of hybrid parameters for one particular transistor set in each of the three configurations.

| CB | CE | CC |
|---|---|---|
| $h_{ib} = 39\ \Omega$ | $h_{ie} = 1900\ \Omega$ | $h_{ic} = 1950\ \Omega$ |
| $h_{rb} = 380 \times 10^{-6}$ | $h_{re} = 575 \times 10^{-6}$ | $h_{rc} = 1$ |
| $h_{fb} = 0.98$ | $h_{fe} = 49$ | $h_{fc} = 50$ |
| $h_{ob} = 0.49\ \mu\text{mho}$ | $h_{oe} = 24\ \mu\text{mho}$ | $h_{oc} = 24.5\ \mu\text{mho}$ |

## 5.2 Parameter Conversion

As mentioned earlier in this chapter, one set of hybrid parameters is generally available on the manufacturer's data sheet. This set is likely to be for the common emitter because this is the most popular configuration. If designers wish a different circuit configuration, they will need to convert the given parameters to make them applicable to the circuit of choice. Also, there is an interrelationship among open-circuit, short-circuit, and hybrid parameters that enables converting from one type of parameter to another.

### 5.2.1 Hybrid Conversion Between Configurations

The following equations are written in mathematical form, but they should be interpreted loosely. In some cases, the quantities separated by the equality sign are not precisely equal to one another. However, the degree of error is so small that it can be safely ignored in circuit design. So these equations are designed for practical application rather than for mathematical precision.

| From CE to CB | From CE to CC |
|---|---|
| $h_{ib} = \dfrac{h_{ie}}{1 + h_{fe}}$ | $h_{ic} = h_{ie}$ |
| $h_{rb} = \dfrac{(h_{ie})(h_{oe})}{1 + h_{fe}} - h_{re}$ | $h_{rc} = 1 - h_{re}$ |
|  | $h_{fc} = 1 + h_{fe}$ |
| $h_{fb} = \dfrac{-h_{fe}}{1 + h_{fe}}$ | $h_{oc} = h_{oe}$ |
| $h_{ob} = \dfrac{h_{oe}}{1 + h_{fe}}$ |  |

| From CB to CE | From CB to CC |
|---|---|
| $h_{ie} = \dfrac{h_{ib}}{1 + h_{fb}}$ | $h_{ic} = \dfrac{h_{ib}}{1 + h_{fb}}$ |
| $h_{re} = \dfrac{(h_{ib})(h_{ob})}{1 + h_{fb}} - h_{rb}$ | $h_{rc} = \dfrac{(h_{ib})(h_{ob})}{1 + h_{fb}} - h_{rb}$ |
| $h_{fe} = \dfrac{h_{fb}}{1 + h_{fb}}$ | $h_{fc} = \dfrac{1}{1 + h_{fb}}$ |
| $h_{oe} = \dfrac{h_{ob}}{1 + h_{fb}}$ | $h_{oc} = \dfrac{h_{ob}}{1 + h_{fb}}$ |

Suppose that a certain transistor has an $h_{ie}$ of 1950 Ω and an $h_{fe}$ of 49 Ω, and it is to be used in a common-base amplifier circuit. The input resistance becomes

$$h_{ib} = \frac{h_{ie}}{1 + h_{fe}}$$

$$= \frac{1950 \ \Omega}{1 + 49}$$

$$= \frac{1950}{50}$$

$$= 39 \ \Omega$$

# Common Problems in Design

The forward current gain becomes

$$h_{fb} = \frac{-h_{fe}}{1 + h_{fe}}$$

$$= \frac{49}{1 + 49}$$

$$= \frac{49}{50}$$

$$= -0.98$$

The common-base parameters for a particular transistor shows an $h_{ob}$ of 0.49 $\mu$mho and an $h_{fb}$ of 0.98. The output conductance for this transistor as a common collector is

$$h_{oc} = \frac{h_{ob}}{1 + h_{fb}}$$

$$= \frac{0.49 \times 10^{-6}}{1 + (-0.98)}$$

$$= 24.4 \ \mu\text{mhos}$$

Forward current gain becomes

$$h_{fc} = \frac{1}{1 + h_{fb}}$$

$$= \frac{1}{1 + (-0.98)}$$

$$= \frac{1}{0.02}$$

$$= 50$$

### 5.2.2 Converting Hybrid to Open Circuit

The following equations are specifically for common-emitter configurations, but they can be applied to any configuration:

$$r_{ie} = \frac{(h_{re})(h_{fe})}{h_{oe}}$$

$$r_{re} = \frac{h_{re}}{h_{oe}}$$

$$r_{fe} = \frac{-h_{fe}}{h_{oe}}$$

$$r_{oe} = \frac{1}{h_{oe}}$$

### 5.2.3 Converting Open Circuit to Hybrid

$$h_{ie} = r_{ie} - \frac{(r_{re})(r_{fe})}{r_{oe}}$$

$$h_{re} = \frac{r_{re}}{r_{oe}}$$

$$h_{fe} = \frac{r_{fe}}{r_{oe}}$$

$$h_{oe} = \frac{1}{r_{oe}}$$

### 5.2.4 Converting Hybrid to Short Circuit

$$g_{ie} = \frac{1}{h_{ie}}$$

$$g_{re} = \frac{-h_{re}}{h_{ie}}$$

$$g_{fe} = \frac{h_{fe}}{h_{ie}}$$

$$g_{oe} = h_{oe} - \frac{(h_{re})(h_{fe})}{h_{ie}}$$

### 5.2.5 Use of Different Hybrid Symbols

Some manufacturers do not use the same symbols for hybrid parameters that have been used in this section. The most frequently encountered exceptions are

$$h_{11}$$
$$h_{12}$$
$$h_{21}$$
$$h_{22}$$

The meanings of these symbols in each configuration are as follows:

# Common Problems in Design

|  | CB | CE | CC |
|---|---|---|---|
| $h_{11}$ = input resistance = | $h_{ib}$ | $h_{ie}$ | $h_{ic}$ |
| $h_{12}$ = reverse voltage gain = | $h_{rb}$ | $h_{re}$ | $h_{rc}$ |
| $h_{21}$ = forward current gain = | $h_{fb}$ | $h_{fe}$ | $h_{fc}$ |
| $h_{22}$ = output conductance = | $h_{ob}$ | $h_{oe}$ | $h_{oc}$ |

Other symbols that are encountered include:

$$\alpha_f$$

$$\mu_r$$

$$\alpha$$

$$\beta$$

$\alpha_f$, ($\alpha$), $h_{21}$, and $h_{fb}$ are four different designations for the same parameter: forward current gain in a common base. The terms $\mu_r$, $h_{12}$, $h_{rb}$, $h_{re}$, and $h_{rc}$ all indicate reverse voltage gain; $h_{21}$, $h_{fe}$, and beta ($\beta$) all indicate forward current gain in a common emitter.

## 5.3 Application of Parameters

Equivalent circuits can be constructed for any configuration. The application of parameters to these equivalent circuits enables calculation of such vital factors as current gain, voltage gain, power gain, input impedance, and output impedance.

### 5.3.1 Common-Base Equivalent Circuit

Since a transistor amplifies both voltage and current, its equivalent is a constant-voltage generator and a constant-current generator. The voltage generator is symbolized with a single circle and the current generator by two interlocking circles. Figure 5-3a is a simplified common-base circuit, and Figure 5-3b is its hybrid-parameter equivalent. The biasing circuits have been omitted.

The input resistance of the device is $h_{ib}$. The circuit input impedance, $Z_i$, is $h_{ib}$, in parallel with $R_E$:

$$Z_i = \frac{h_{ib} \times R_E}{h_{ib} + R_E}$$

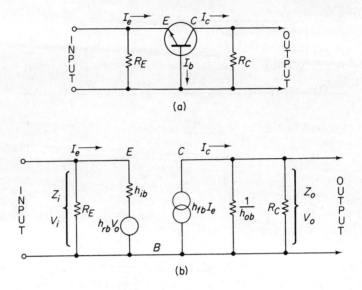

Figure 5-3. Common-Base Circuit and Its Equivalent

Since $R_E$ is usually more than 10 times the value of $h_{ib}$, the input impedance is approximately equal to $h_{ib}$:

$$Z_i = h_{ib}$$

The voltage generator is producing a voltage equal to the output voltage, $V_o$, multiplied by the reverse voltage gain, $h_{rb}$.

The current generator is producing a current equal to the emitter current, $I_e$, multiplied by the forward current transfer ratio, $h_{fb}$. This current is $I_c$; and $I_c$ divides between the ac collector resistance and the collector load resistance, $R_C$. The ac collector resistance is given by the reciprocal of the output conductance, $1/h_{ob}$.

The output impedance is $R_C$, in parallel with $1/h_{ob}$:

$$Z_o = \frac{R_C \times (1/h_{ob})}{R_C + (1/h_{ob})}$$

Since $R_C$ is usually less than one tenth of the value of $1/h_{ob}$, the output impedance is approximately equal to $R_C$:

$$Z_o = R_C$$

# Common Problems in Design

The voltage gain is the output voltage divided by the input voltage:

$$A_v = \frac{I_c R_C}{I_e h_{ib}}$$

This equation can be reduced to

$$A_v = \frac{h_{fb} R_C}{h_{ib}}$$

which is forward-current transfer ratio times $R_C$ divided by the input resistance of the device.

Current gain is output current divided by input current. In the common base, current gain, $A_i$, is always less than unity, and it is approximately equal to alpha, which is equal to $h_{fb}$:

$$A_i = h_{fb} = \alpha$$

Power gain is current gain multiplied by voltage gain. This is

$$G_p = A_v A_i$$

In terms of voltage- and current-gain equations, power gain can be expressed as

$$G_p = \frac{(h_{fb})^2 R_C}{h_{ib}}$$

The common base provides gain at both voltage and power, but the current gain is less than unity. It has a low input impedance and a high output impedance.

### 5.3.2 Common-Emitter Equivalent Circuit

The common emitter is the most used of the three basic circuits. It provides gain of voltage, current, and power. The voltage has a phase inversion through the common emitter; output is 180 degrees out of phase with the input. Figure 5-4a is a simplified common-emitter circuit, and Figure 5-4b is its hybrid-parameter equivalent circuit.

The input resistance of a common emitter is $h_{ie}$. The circuit input impedance is $h_{ie}$ in parallel with $R_B$:

$$Z_i = \frac{h_{ie} R_B}{h_{ie} + R_B}$$

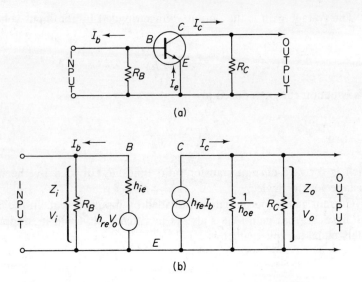

Figure 5-4. Common-Emitter Circuit and Its Equivalent

The input resistance, $h_{ie}$, ranges from 500 Ω to 1500 Ω. The value of $R_B$ is likely to be on the order of 1 million Ω. This means that $h_{ie}$ is less than one tenth of $R_B$, so $Z_i$ is approximately equal to $h_{ie}$:

$$Z_i = h_{ie}$$

The voltage generator is producing a voltage equal to the output voltage, $V_o$, multiplied by the reverse voltage gain, $h_{re}$. The current generator is producing a current equal to the input current, $I_b$, multiplied by the forward-current transfer ratio, $h_{fe}$. This current is the same as $I_c$, and it is divided between the output resistance (ac collector resistance) of the transistor, $1/h_{oe}$, and the collector load resistance, $R_C$. Since $h_{oe}$ is the output conductance, $1/h_{oe}$ is the output resistance.

The output impedance of the circuit is $R_C$, in parallel with $1/h_{oe}$:

$$Z_o = \frac{R_C \times (1/h_{oe})}{R_C + (1/h_{oe})}$$

The ac output resistance, $1/h_{oe}$, ranges from 30 kΩ to more than 1 million Ω. In most cases, $R_C$ is much smaller than $1/h_{oe}$. This causes $Z_o$ to be approximately equal to $R_C$:

$$Z_o = R_C$$

# Common Problems in Design

The voltage gain is output voltage divided by input voltage:

$$A_v = \frac{I_c R_C}{I_b h_{ie}}$$

This gain equation may be rewritten as

$$A_v = \frac{h_{fe} R_C}{h_{ie}}$$

This equation is forward-current transfer ratio times $R_C$ divided by the ac input resistance. It should be remembered that this output voltage is 180 degrees out of phase with the input voltage.

Current gain is output current divided by input current. In the common emitter, current gain, $A_i$, is approximately equal to beta, which is equal to $h_{fe}$:

$$A_i = h_{fe} = \beta$$

Power gain is current gain multiplied by voltage gain:

$$G_p = A_v A_i$$

In terms of previous equations, power gain can be expressed as

$$G_p = \frac{(h_{fe})^2 R_C}{h_{ie}}$$

The common emitter is very popular as a voltage amplifier. It provides gain of current, voltage, and power. It has a relatively high input impedance and a high output impedance.

### 5.3.3 Common-Collector Equivalent Circuit

The common collector is a very handy circuit when it becomes necessary to match a high impedance to a low impedance. It provides a current gain and a power gain, but its voltage gain is always less than unity. Figure 5-5a is a simplified common-collector circuit, and Figure 5-5b is its hybrid-parameter equivalent.

The input resistance of a common collector is $h_{ic}$. The circuit input impedance is $h_{ic}$, in parallel with $R_B$:

$$Z_i = \frac{h_{ic} R_B}{h_{ic} + R_B}$$

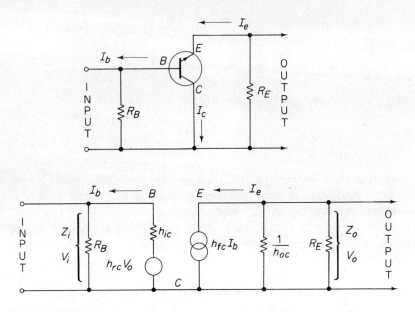

Figure 5-5. Common-Collector Circuit and Its Equivalent

The input resistance, $h_{ic}$, is usually very large. It may range from 20 k$\Omega$ to more than 1 million $\Omega$. So regardless of the value of $R_B$, $h_{ic}$ is a significant factor in the circuit input impedance.

The voltage generator is producing a voltage equal to the output voltage multiplied by the reverse voltage gain, $h_{rc}$. The current generator is producing a current equal to the input current, $I_b$, multiplied by the forward-current transfer ratio, $h_{fc}$. This current is the same as $I_e$, and it divides between the output resistance of the device, $1/h_{oc}$, and the emitter load resistor, $R_E$. Since $h_{oc}$ is the output conductance, $1/h_{oc}$ must be the output resistance.

The output impedance of the circuit is $R_E$, in parallel with $1/h_{oc}$:

$$Z_o = \frac{R_E \times (1/h_{oc})}{R_E + (1/h_{oc})}$$

This equation is an approximation because any variation in output voltage has a significant effect on the input circuit. The output resistance of the device, $1/h_{oc}$, is something less than 1000 $\Omega$. Since $R_E$ is usually more than 10 times the value of $1/h_{oc}$,

$$Z_o = \frac{1}{h_{oc}}$$

# Common Problems in Design

The current gain of the transistor is equal to output current, $I_e$, divided by input current, $I_b$. This is also equal to $h_{fc}$ and to beta $+1$:

$$A_i = \frac{I_e}{I_b} = h_{fc} = \beta + 1$$

This device current gain is larger than the circuit current gain because the input current divides between $R_B$ and $h_{ic}$. The circuit current gain is approximately

$$A_i = \frac{h_{fc} R_B}{h_{ic} + R_B}$$

The power gain is voltage gain multiplied by current gain. Since the voltage gain is always slightly less than unity, the power gain is approximately equal to current gain:

$$G_p = A_v \times A_i = A_i$$

## 5.4 Effects of Frequency on Parameters

Most data sheets specify the parameters at a given frequency. This frequency standard is generally 1 kHz. One design problem is estimating these parameters when the circuit is operated at some other frequency. Current gain is a parameter that appears on nearly all data sheets. This current gain is alpha or $h_{fb}$ for a common base, and beta or $h_{fe}$ for a common emitter. In both circuits the forward current gain is affected by frequency.

### 5.4.1 Current Gain Versus Frequency

Sometimes a data sheet will list $h_{fbo}$ for a common base and $h_{fco}$ for a common emitter. These designators specify the value of forward current gain at a frequency of 1 kHz. Closely related to these parameters is the current-gain cutoff frequency. The cutoff frequency for a common base is specified as $f_{ab}$; for a common emitter it is $f_{ae}$. The cutoff frequency is that frequency which causes the current gain to decrease 3 dB from the standard frequency value. The current gain–frequency relationship is illustrated by the graph in Figure 5-6. The cutoff frequency, $f_a$, is the point where current gain is 3 dB down from the value of 1 kHz. Current gain at this frequency is 0.707 of the gain at 1 kHz. After the cutoff frequency is reached, the gain drops off very rapidly.

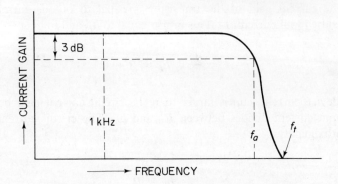

Figure 5-6. Current Gain Versus Frequency

### 5.4.2 Gain–Bandwidth Product

Some data sheets list another frequency parameter for common emitters. It is called a *gain–bandwidth product* and is designated as $f_T$, as shown on the graph:

$$f_T = f_{ae} \times h_{feo}$$

This gain–bandwidth product is the frequency where current gain drops to 0 dB. The current gain at 0 dB is still equal to unity. So $f_T$ is not necessarily the maximum operating frequency.

### 5.4.3 Maximum Operating Frequency

The *maximum operating frequency* is the frequency that causes the power gain to drop to unity, and it is designated as $f_{\max}$. This $f_{\max}$ is not always identified in this manner. Some data sheets list one power gain for 1 kHz and another at some frequency above $f_{ae}$. Any frequency above $f_{ae}$ may be used to calculate $f_{\max}$ by using the following equation:

$$f_{\max} = f\sqrt{G_p}$$

where $f$ is the specified frequency above $f_{ae}$ and $G_p$ is dB converted to power-gain magnitude.

## 5.5 Effects of Temperature on Parameters

Parameters listed on data sheets are true for a specific temperature. This specified, or understood, temperature is 25°C. Of course, transistors seldom operate at room temperature, and when temperature changes, the parameters change. And when some parameters change, this causes further temperature changes. An example of this interaction is thermal runaway.

### 5.5.1 Current Gain

A drastic change in temperature causes a drastic change in current gain. The germanium transistor is more sensitive to temperature changes than silicon transistors. In both types, current gain increases as temperature increases. A germanium transistor has about twice the gain at 100°C that it has at 25°C. At 175°C, the silicon transistor has about twice its rated current gain.

The recommended maximum operating temperature for a silicon transistor is 200°C, and for germanium it is 100°C. Operation above these temperatures is likely to destroy the transistor.

### 5.5.2 Power Dissipation

Power dissipation is also directly associated with temperature. The maximum power dissipation is specified on the data sheet. The power rating is sometimes given in terms of 25°C with a derating factor as temperature rises. The rated maximum power dissipation must not be exceeded. In fact, maximum power should be avoided by a safe margin. As temperature rises, current increases, temperature increases, the transistor gets hotter, and more power must be dissipated.

### 5.5.3 Leakage Current

Leakage current between collector and base, $I_{cbo}$, is directly proportional to temperature. This $I_{cbo}$ is a minority carrier current, and reverse bias for the collector–base junction is forward bias for the minority carriers. As temperature rises, more minority carriers are released, $I_{cbo}$ increases, and collector current rises. The direction of base current is opposite to the direction of $I_{cbo}$. The canceling effect between $I_b$ and $I_{cbo}$ hinders the ability of $I_b$ in controlling the collector current.

## 5.6 Use of Heat Sinks

The manufacturer packages his transistors in cases that aid in dissipating heat. These cases can dissipate heat according to their physical size and the type of metal used in their construction.

When additional heat dissipation is required, heat sinks must be used. There are various types of heat sinks: metal washers, direct chassis contact, and metal fin-like flanges. Some devices might even be liquid-cooled.

Commercial heat sinks are rated in terms of thermal resistance expressed as degrees celsius per watt of power. These heat sinks are manufactured to fit specified transistors. The manufacturers' recommendations should be followed.

## 5.7 Thermal Circuits

*Thermal circuit* is a term used to describe the process of heat generation and transfer. The thermal circuit is best explained by comparing it to an electrical circuit.

### 5.7.1 Electrical Analogy

Thermal resistance is opposition to the transfer of heat. This term came about by comparing thermal action to resistance in an electrical circuit. Temperature and junction power can also be compared to electrical properties. This analogy is illustrated in Figure 5-7.

In this drawing, the following factors are analogous to each other:

$$\text{current} \approx \text{junction power}$$
$$\text{voltage} \approx \text{temperature}$$
$$\text{resistance} \approx \text{thermal resistance}$$

The total thermal resistance, $\Theta_{JA}$, is composed of two thermal resistors in series: junction to case, $\Theta_{JC}$, and case to ambient air, $\Theta_{CA}$:

$$\Theta_{JA} = \Theta_{JC} + \Theta_{CA}$$

Thermal resistance is expressed in degrees celsius per watt of power.

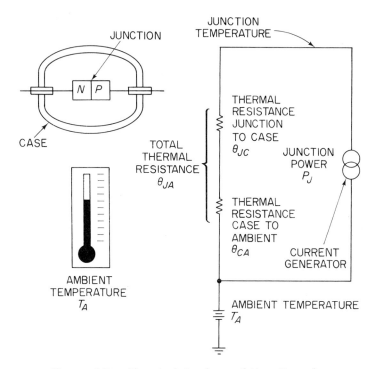

Figure 5-7. Electrical Analogy of Heat Transfer

The junction temperature, $T_J$, is the product of thermal resistance times junction power:

$$T_J = \Theta_{JA} \times P_J$$

The junction temperature is the rise in temperature above the ambient temperature. The objective is to keep $T_J = T_A$ for the most efficient operation.

A heat sink enlarges the cooling surface and reduces the thermal resistance. When a heat sink is used, the case to ambient resistance, $\Theta_{CA}$, drops from the equation and is replaced by two others. The two new thermal resistances are case to sink, $\Theta_{CS}$, and sink to ambient, $\Theta_{SA}$. The sum of the two new resistors is always less than the case to ambient, $\Theta_{CA}$, resistor that they replaced:

$$\Theta_{CA} > \Theta_{CS} + \Theta_{SA}$$

Thus the addition of a heat sink enables more power per degree of junction temperature. The total thermal resistance with a heat sink is

$$\Theta_{JA} = \Theta_{JC} + \Theta_{CS} + \Theta_{SA}$$

These thermal resistors and their relation to the transistor are illustrated in Figure 5-8.

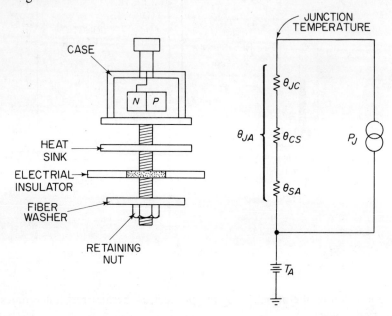

Figure 5-8. Thermal Circuit with a Heat Sink

### 5.7.2 Maximum Power

The maximum allowable operating temperature, $T_{J(MAX)}$, is still a power-limiting factor. The heat sink does not alter this fact. However, with a heat sink, the semiconductor can handle considerably more power before the junction reaches this critical temperature. The maximum power can be calculated from the values of the thermal factors:

$$P_{J(MAX)} = \frac{T_{J(MAX)} - T_A}{\Theta_{JA}}$$

In relation to the thermal resistors,

$$\Theta_{JA} = \frac{T_{J(MAX)} - T_A}{P_J}$$

## Common Problems in Design

Therefore,

$$\Theta_{JC} + \Theta_{SC} + \Theta_{SA} = \frac{T_{J(MAX)} - T_A}{P_J}$$

*Sample Problem:* What is the maximum safe power at an ambient temperature of 40°C for a diode with the following specifications?

$$T_{J(MAX)} = 150°C$$
$$\Theta_{JA} = 6°C/W$$

*Solution:*

$$P_{J(MAX)} = \frac{T_{J(MAX)} - T_A}{\Theta_{JA}}$$
$$= \frac{150 - 40}{6}$$
$$= 18.83 \text{ W}$$

In some cases, a maximum case temperature, $T_{C(MAX)}$, is specified rather than a maximum junction temperature. The maximum power is then equal to case temperature minus ambient temperature divided by total thermal resistance. The equation then becomes

$$P_{J(MAX)} = \frac{T_{C(MAX)} - T_A}{\Theta_{JA}}$$

When the specifications are

$$T_{C(MAX)} = 150°C$$
$$\Theta_{JA} = 6°C/W$$

then at an ambient temperature of 50°C the maximum safe power is

$$P_{J(MAX)} = \frac{T_{C(MAX)} - T_A}{\Theta_{JA}}$$
$$= \frac{150 - 50}{6}$$
$$= 16.67 \text{ W}$$

Since all data sheets are not the same, it is difficult to determine the arrangement of the equations to fit every need. However, slight manipulation of these equations should produce any quantity that needs to be calculated. For proper operation, the quantity $\Theta_{JA}$ must always be less than $T_{J(MAX)}/P_{J(MAX)}$. In the case of a bipolar transistor,

$$P_J = I_C \times V_C$$

Remember that total $I_C$ must include $I_{CBO}$.

## 5.8 Test Equipment

Having the proper test equipment on hand when needed is always a problem in design. However, with a bit of planning, this need not be a serious problem. Actually only five pieces of test equipment are required in general design projects:

Transistor tester.
Multimeter.
Signal generator.
Oscilloscope.
Regulated power supply.

### 5.8.1 Transistor Tester

The transistor tester is a much needed and often used device. It makes the difference between knowing a transistor is good and just assuming that it is good. The transistor tester should be capable of the following tests as a minimum specification:

Diodes.
Rectifier.
Signal (dynamic).
Zener.
Bipolar transistors.
$I_{cbo}$.
$I_{ceo}$.
Alpha.
Beta.
Punch-through voltage.
Hybrid parameters.
Field-effect transistors.
Transconductance.
Leakage.

A lot of time and trouble is saved by a transistor tester capable of making these tests with the transistor in the circuit. This makes it unnecessary to disconnect the component each time before testing.

### 5.8.2 Multimeter

A multimeter is needed constantly. In fact, many situations arise that require two multimeters. In solid-state design, the multimeter must be capable of measuring very small values of input current and voltage. So it should have one or more scales that a full-scale deflection is no more than 50 mV. The same meter may need to measure supply voltages and currents of relatively large quantities. So the meter should have a variety of range settings. Since signals are ac in nature, a meter capable of reading peak-to-peak value will save a lot of calculating. The multimeter should be of the electronic type.

### 5.8.3 Signal Generator

The signal generator is needed to supply the circuit with a test signal. The frequency response of the signal generator must at least match or exceed the response of the circuit design. The output of the signal generator should be a low impedance. The output voltage should be variable in precise steps from 0 to 10 V.

### 5.8.4 Oscilloscope

The oscilloscope should have a triggered sweep to enable synchronization with an external signal. Without this feature, a lot of time is wasted trying to have a signal stand still long enough for adequate observation. A differential input to the oscilloscope will minimize the pickup of stray noise and simplify the observation of small signals. A *dual trace* is also a highly desirable feature. The dual trace enables observation of both input and output signals at the same time. It also enables the comparison of a doubtful waveshape with another wave of known accuracy.

### 5.8.5 Regulated Power Supply

The regulated power supply is more a part of the circuit under design than it is a piece of test equipment. It should be capable of supplying an adjustable direct voltage output from 0 to 50 V, either positive or negative. This power supply is to furnish both input and output bias voltages for the circuit. Since shorts often occur in experimental circuits, the power supply should have built-in protection against shorts.

## 5.9  Spare Parts

It is not necessary to maintain a large inventory of spare parts to facilitate efficient design. Standard parts, and some precision devices, are readily available in electronic supply stores. However, time and frustration can be saved by keeping a limited stock of often-needed parts. The spare parts should be of three types:

Decade boxes.
Fixed components.
Variable components.

### 5.9.1  Decade Boxes

The *decade boxes* of resistors, capacitors, and inductors can save a lot of time and work since all design work is, to some extent, based on trial and error. As a minimum stock, a designer should have the following decade boxes in the quantity indicated:

Resistor, 4.
Capacitor, 2.
Inductor, 2.

### 5.9.2  Fixed Components

A ready supply of *fixed components,* consisting of a few standard value resistors and capacitors, is essential. A little planning will indicate the appropriate quantity and most likely values to include in this stock. In addition, there should be several junction diodes, a few Zener diodes, and at least two spares of the semiconductor device featured in the design circuit.

### 5.9.3  Variable Components

*Variable components* tend to be expensive, but a few variable resistors and capacitors may be a wise investment. In most cases, a variable component can be replaced with the nearest standard value in the final circuit. However, the variable components are the easiest method of obtaining peak performance by compensating for various types of circuit interaction.

SECTION THREE

# AUDIO-FREQUENCY CIRCUITS

A large majority of audio-frequency circuits are composed of amplifiers. These amplifiers are of two types: voltage amplifiers and power amplifiers. In most cases, the amplifying device is the designer's choice of either a bipolar or a field-effect transistor. This section treats both voltage and power amplifiers, and the examples include both types of transistor. Noise and feedback in audio-frequency amplifiers are two problems of sufficient significance to warrant special discussions.

Apart from amplifiers, the audio-frequency circuits are concerned with microphones, speakers, and audio oscillators. The audio oscillator is merely a special application of an amplifier. This oscillator is covered in Chapter Fourteen. The design of microphones and speakers is outside the scope of this book.

CHAPTER SIX

# Bipolar Voltage Amplifiers

An audio-frequency (AF) *voltage amplifier* must be designed to deliver to the load an enlarged version of the input signal. This must be a reasonably high fidelity output. So the circuit must be designed for a minimum of distortion. Audio-voltage amplifiers may be either small-signal or large-signal amplifiers, so they are analyzed by both equivalent circuits and load lines. In most cases, a voltage amplifier produces a positive gain in both current and power as well as voltage, but the signal voltage waveshape is the primary consideration.

## 6.1 Basic Bipolar Amplifier

The basic bipolar AF voltage amplifier is generally a common-emitter configuration, but in some special cases, common-base circuits are used. It must be remembered that the common base does not amplify current, and the common collector does not amplify voltage. The basic amplifier is little different from the examples previously discussed. The main difference is in the limiting nature of the specifications. The specifications dictate many of the circuit components, and practical design is a matter of getting the optimum performance under the prescribed circumstances.

### 6.1.1 Model Circuit

The first step is to design the model circuit on paper and to indicate the restricting specifications. An AF voltage amplifier circuit with moderately high fidelity is similar to that in Figure 6-1.

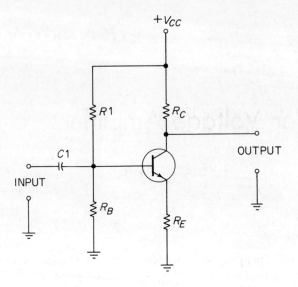

Figure 6-1. Model Bipolar Circuit

This circuit utilizes an *NPN* transistor with *RC* coupling on the input. The output is taken across $R_C$ and may be *RC*-coupled to the input of the next stage. Since this is a circuit that will be required to handle only audio frequencies, only the low-frequency limit of the coupling capacitor needs to be considered. The emitter resistor, $R_E$, provides degenerative feedback to cause a good reproduction of the input signal and to improve the stability of the transistor.

Resistors $R1$ and $R_B$ form a voltage divider for the base-to-emitter bias. The bias is the difference between the voltage drop across $R_E$ and the voltage drop across $R_B$. The relative values of $R1$ and $R_B$ are selected to produce the required value of static collector current.

### 6.1.2 Estimating Component Values

For the moment, the circuit component values will be estimated from either specifications or other limiting conditions. The available voltage from the power supply is generally a fixed value. Suppose that it is +20-V-regulated. This establishes the value of $V_{CC}$ and sets the limit for the maximum swing of collector voltage.

Since maximum power transfer is always a desirable feature, the input impedance should match the output impedance of the previous stage, and the output impedance should match the input impedance of the following stage. A perfect match is difficult to obtain, but an effort must be made to come as close as reasonably possible. For trial purposes, the input impedance of this stage may be considered as approximately the same as the value of $R_B$. So if the output impedance of the previous stage is 2000 Ω, a 2000-Ω resistor should be used for $R_B$. The output impedance of this stage is approximately the same as $R_C$. So if the input impedance of the next stage is 4000 Ω, $R_C$ should be 4000 Ω.

The feedback gain and stability factors should now be considered. Voltage gain is roughly $R_C$ divided by $R_E$. Assuming that a minimum gain of 10 is required, $R_C$ should be 10 times the value of $R_E$. Since $R_C$ is 4000 Ω, $R_E$ must be 400 Ω or less. The stability factor is $R_B$ divided by $R_E$. The higher the stability, the lower the gain. So a compromise must be reached. A stability factor of 20 is high gain with low stability; a factor of 5 is low gain and high stability. For an AF voltage amplifier, 10 is a desirable stability factor. A factor of 10 will allow optimum gain with optimum stability. The value of $R_E$ is thus established as one tenth of $R_B$. Since $R_B$ is 2000 Ω, $R_E$ becomes 200 Ω.

Now the biasing circuit can be completed. The value of $R1$ sets the bias and determines the static current–voltage conditions. For maximum, undistorted output, the static $V_C$ should be set to about half the value of $V_{CC}$. Since $V_{CC}$ is +20 V, the static $V_C$ should be +10 V. The value of $I_C$ that will produce this +10 V of $V_C$ is calculated by

$$I_C = \frac{V_C}{R_C}$$
$$= \frac{10 \text{ V}}{4 \text{ k}\Omega}$$
$$= 2.5 \text{ mA}$$

It seems that the value of $R1$ must provide a bias that produces an $I_C$ of 2.5 mA. But there is another factor to be considered. The input bias voltage must still be some positive value at the negative peak of the input signal. So if the input has a peak value of 0.2 V, the voltage drop across $R_B$ should be set to at least 0.7 V for a germanium transistor and at 1 V for a silicon transistor.

The easiest way to set the proper bias is to use a variable resistor for $R1$. It can then be adjusted until $V_B$ is of the desired value. In the final circuit, $R1$ can be measured and replaced by a fixed resistor of comparable size. The size can be estimated if a fixed resistor must be used. With an $I_C$ of 2.5 mA, the $I_B$ is probably about 0.2 mA. This 0.2 mA through $R1$, plus current through $R_B$, must drop all of $V_{CC}$ except the

1 V desired across $R_B$ (assuming a silicon transistor). In this case, current through $R_B$ is

$$IR_B = \frac{1 \text{ V}}{2 \text{ k}\Omega}$$
$$= 0.5 \text{ mA}$$

Current through $R1$ then must be

$$I_{R1} = I_{R_B} + I_B$$
$$= 0.5 \text{ mA} + 0.2 \text{ mA}$$
$$= 0.7 \text{ mA}$$

This 0.7 mA must drop the remaining 19 V of $V_{CC}$ across $R1$. So

$$R1 = \frac{19 \text{ V}}{0.7 \text{ mA}}$$
$$= 27 \text{ k}\Omega$$

A decade box can be used to substitute the nearest available resistor, and $VR_B$ can be measured. Other values of resistance can be switched in until the value of $VR_B$ approaches the necessary 1 V.

Now for the value of the coupling capacitor. Only the low-frequency cutoff for this capacitor needs to be considered. The low-frequency cutoff should be near the bottom of the audio band. The cutoff frequency should be considered as that frequency which will cause the value of $X_C$ to be half the ohmic value of $R_B$. With the value of $R_B$ being 2000 $\Omega$, and the low-frequency cutoff being set for 50 Hz, the capacitance is

$$C = \frac{1}{2\pi f X_C}$$
$$= \frac{0.159}{50 \times 1 \times 10^3}$$
$$= \frac{0.159 \times 10^{-6}}{0.05}$$
$$= 3.2 \ \mu\text{F}$$

Since this is the minimum acceptable capacitance, the next higher standard capacitance will serve very well.

## Bipolar Voltage Amplifiers

The transistor is not a critical item. About the only characteristic that could cause a problem would be a beta value less than the required circuit gain. Since minimum circuit gain is 10, any beta in excess of 10 should serve the purpose. The beta for most transistors exceeds this modest value. Suppose that the selected transistor is an *NPN* silicon, and it has the following characteristics:

$$h_{ie} = 2000 \ \Omega$$

$$h_{fe} = 16$$

$$h_{re} = 500 \times 10^{-6}$$

$$h_{oe} = 24 \ \mu\text{mhos}$$

Now the circuit of Figure 6-1 can have the values inserted as shown in Figure 6-2. A circuit exactly like this would probably do a reasonable job of meeting the specifications. But this is only a paper model to work from. Now the model circuit needs to be analyzed. Chances are very good that some refinements can be made before the actual test circuit is built.

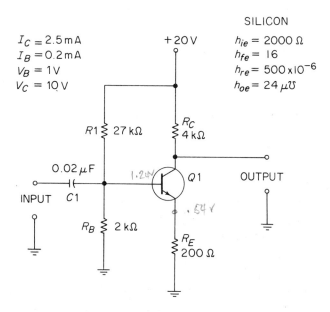

Figure 6-2. Model Circuit with Trial Components

## 6.2 Hybrid Equivalent Analysis

The model circuit has all the necessary information for constructing the hybrid-parameter equivalent circuit. This circuit should be constructed and analyzed as the next step in the design process.

### 6.2.1 Constructing the Circuit

The procedure for constructing the hybrid equivalent circuit is discussed in Chapter Five. This is a common-emitter configuration, so the equivalent circuit will be similar to that in Figure 6-3. When the values from the model circuit are substituted onto this example, the values of gain and impedance can be calculated.

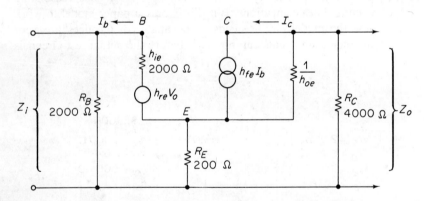

Figure 6-3. Hybrid Equivalent of Figure 6-2

### 6.2.2 Input Impedance

The voltage, $h_{re}V_o$, is the portion of the output voltage that feeds back into the input circuit. This feedback voltage in a common emitter is very small when compared to the voltage drop across $h_{ie}$. In fact, $h_{re}V_o$ is so small that it can be completely ignored. The input resistance for the transistor then becomes:

$$R_{in} = h_{ie} + R_E(h_{fe} + 1)$$
$$= 2000 + 200(16 + 1)$$
$$= 2000 + 3400$$
$$= 5400 \ \Omega$$

The circuit input impedance is $R_B$, in parallel with $R_{in}$:

$$Z_i = \frac{R_B \times R_{in}}{R_B + R_{in}}$$

$$= \frac{2000 \times 5400}{2000 + 5400}$$

$$= \frac{108{,}000}{74}$$

$$= 1460 \;\Omega$$

This impedance is about 25 percent too low. It should be 2000 $\Omega$, to match the output impedance of the previous stage. A 25 percent mismatch is not too bad if the remainder of the circuit checks out. However, the value of $R_B$ could be raised to 3000 $\Omega$. This increase in $R_B$ would provide a nearly perfect impedance match for the input, but it might necessitate other changes.

### 6.2.3 Output Impedance

The output resistance of the transistor is $1/h_{oe}$. The 200-$\Omega$ $R_E$ is too small to consider. Since $h_{oe}$ is 24 $\mu$mhos:

$$R_{\text{out}} = \frac{1}{h_{oe}}$$

$$= \frac{1}{24 \times 10^{-6}}$$

$$= \frac{1 \times 10^6}{24}$$

$$= 41{,}666 \;\Omega$$

The output impedance of the circuit is this output resistance in parallel with $R_C$:

$$Z_o = \frac{R_{\text{out}} \times R_C}{R_{\text{out}} + R_C}$$

Since $R_{\text{out}}$ is more than 10 times the value of $R_C$, the output impedance is approximately the same as $R_C$. So the 4000-$\Omega$ value for $R_C$ is the proper value, and it matches the impedance of the next stage.

### 6.2.4 Voltage Gain

Usually the product of $R_E(h_{fe} + 1)$ is more than 10 times the value of $h_{ie}$. When this is true, the voltage gain is

$$A_v = \frac{R_C}{R_E}$$

In the model circuit under consideration, this 10:1 ratio is *not* true. So, in this case, other factors must be considered. The voltage gain is calculated by

$$A_V = \frac{h_{fe}R_C}{h_{ie} + R_E(h_{fe} + 1)}$$

$$= \frac{16 \times 4000}{2000 + 200(16 + 1)}$$

$$= \frac{64{,}000}{5400}$$

$$= 12$$

This is, of course, a very low voltage gain. It is not uncommon to have a common emitter with a voltage gain in excess of 300. But this gain of 12 exceeds the specifications and satisfies the objectives very well.

### 6.2.5 Current and Power Gain

Since this is a voltage amplifier, the gain of current and power are of only academic interest. The current gain is

$$A_i = \frac{h_{fe}R_B}{R_B + Z_i}$$

$$= \frac{16 \times 2000}{2000 + 1460}$$

$$= 9.25$$

The power gain is

$$G_p = A_i \times A_v$$

$$= 9.25 \times 12$$

$$= 111$$

# Bipolar Voltage Amplifiers

### 6.2.6 Summary of the Hybrid Analysis

The results of the hybrid-equivalent circuit analysis are very favorable. They show that the selected component values meet all the specifications. The one area that is borderline is the matching of the input impedance. The 25 percent mismatch will result in some loss of power but not enough to cause concern.

If greater stability is desired, both $R_E$ and $R_B$ values could be increased. But if this is done, a transistor with a higher value of $h_{fe}$ would have to be used in order to maintain a voltage gain in excess of 10.

## 6.3 Deriving Logical Equations

All the previous equations could be memorized, and as the example indicates, they serve the purpose very well. However, a little logical reasoning will produce results almost as accurate as the involved calculations. After all, these equations were produced from logic.

### 6.3.1 Inside the Transistor

It is logical that a transistor has an internal resistance associated with each element. These resistances are illustrated in Figure 6-4. These are ac resistances in series with the transistor elements as indicated.

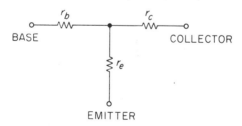

Figure 6-4. Internal Resistances

The base resistance, $r_b$, ranges from 200 to 800 Ω. The emitter resistance is $r_e$, and it is a very small resistance. The value of $r_e$ can be estimated by dividing 26 by the static value of emitter current in milliamperes. The collector resistance, $r_c$, ranges from 10 kΩ to 100 kΩ for either a common emitter or a common collector. For the common base, the value of $r_c$ increases because it is in a different position. The common base $r_c$ is 10 kΩ to 100 kΩ multiplied by beta.

Now imagine a current generator in parallel with $r_c$, as illustrated in Figure 6-5.

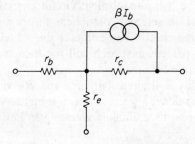

Figure 6-5.  Internal Resistance with a Current Generator

### 6.3.2  Effects of the External Circuit

The external circuit modifies the characteristics of the transistor. Just how much modification is determined by

The circuit configuration.
The circuit component values.

Part of each configuration is common to both input and output. For example, in the common-emitter configuration, the emitter circuit is part of the input circuit and part of the output circuit. This arrangement is obvious in the simplified circuit of Figure 6-6.

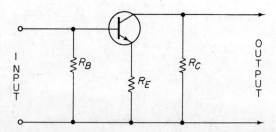

Figure 6-6.  Simplified Common-Emitter (CE) Diagram

Looking into the input side, $R_B$ is in parallel with $R_E$. Looking into the output side, $R_C$ is in parallel with $R_E$. When the transistor is replaced by its equivalent resistance diagram, the relationship of internal and external circuits become just as obvious. Figure 6-7 is a combination of Figures 6-5 and 6-6. It is easy to see that the input circuit consists of $R_B$ in parallel with $r_b + r_e + R_E$. By the same examination, it

## Bipolar Voltage Amplifiers

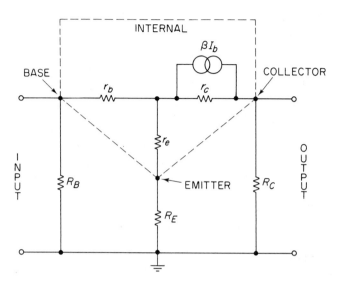

Figure 6-7. Relationship of Internal and External CE Circuits

can be seen that the output circuit consists of $R_C$ in parallel with $r_c + r_e + R_E$. The values of the common part of the circuit are modified by the beta of the transistor. The input resistance from base to ground is

$$R_{\text{in}} = r_b + \beta(r_e + R_E)$$

Resistance $r_b$ is so small in comparison to $\beta(r_e + R_E)$ that $r_b$ has very little effect on $R_{\text{in}}$. The input impedance as seen from the previous stage is $R_B$ in parallel with $R_{\text{in}}$:

$$Z_i = \frac{R_B R_{\text{in}}}{R_B + R_{\text{in}}}$$

In the output circuit, the output resistance from collector to ground is

$$R_{\text{out}} = r_c + \beta(r_e + R_E)$$

The output-circuit impedance is $R_C$ in parallel with $R_{\text{out}}$:

$$Z_o = \frac{R_C R_{\text{out}}}{R_C + R_{\text{out}}}$$

When $R_{out}$ is more than 10 times $R_C$, the output impedance is considered to be the same as $R_C$. This is the usual case with a common emitter.

$$A_v \approx \frac{R_C}{R_E}$$

$$A_i \approx \beta$$

$$G_p \approx \frac{\beta R_C}{R_E}$$

### 6.3.3 Common Collector

The common-collector (CC) internal and external circuits are illustrated in Figure 6-8. In this configuration $r_c$ is part of both input and output circuits, but note that $r_e + R_E$ are in parallel with $r_c$. The sum of $r_e + R_E$ is so much smaller than $r_c$ that it has no significant effect on either input or output impedance. $R_{in}$ is the same as for a CE circuit:

$$R_{in} = r_b + \beta(r_e + R_E)$$

By the same token, the base circuit becomes part of the output circuit:

$$R_{out} = r_e + \beta r_b$$

$$A_v \approx 1$$

$$A_i \approx \beta + 1$$

$$G_p \approx \beta + 1$$

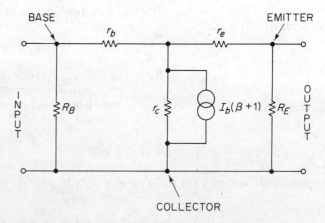

Figure 6-8. Internal and External CC Circuits

### 6.3.4 Common Base

The relationships of internal and external resistances of a common-base (CB) circuit are illustrated in Figure 6-9. Now the sum of $r_b + R_B$ is common to both input and output circuits.

$$R_{\text{in}} = r_e + \frac{r_b + R_B}{\beta}$$

The output impedance of the previous stage now plays a role in both output impedance and gain.

$$R_{\text{out}} \approx R_C$$

$$A_v \approx \frac{R_C}{Z_{\text{gen}}}$$

where $Z_{\text{gen}}$ = output impedance of the previous stage.

$$A_i \approx \alpha$$

$$G_p \approx \alpha A_v$$

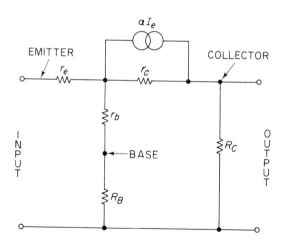

Figure 6-9. Internal and External CB Circuits

## 6.4 Load-Line Analysis

The load line is another convenient way to analyze how a transistor functions in a particular circuit. A family of characteristic curves for the particular transistor is essential to this analysis. Since these curves are unavailable for a common-collector configuration, the load lines apply only to CE and CB circuits. Since this chapter concerns voltage amplifiers, the common collector is of no value anyway; it does not amplify voltage. It is a simple matter to construct a load line with a minimum of knowledge about both the transistor and the circuit. Simplified audio circuits are used to illustrate the use of load lines for both CB and CE configurations.

### 6.4.1 Common-Base Amplifier

Figure 6-10 is a model circuit that needs some paper analysis before the test circuit is built. Coupling and feedback have been eliminated for simplicity. For the moment, the supply voltage and the value of $R_C$ are the only known quantities.

$$R_C = 5 \text{ k}\Omega$$
$$V_{CC} = -20 \text{ V}$$

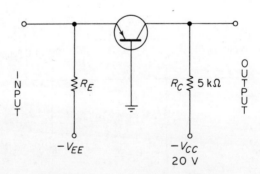

Figure 6-10. Model CB Circuit

The transistor is a *PNP*, and all other known quantities are contained on the matching characteristics graph. The graph for this transistor is shown in Figure 6-11.

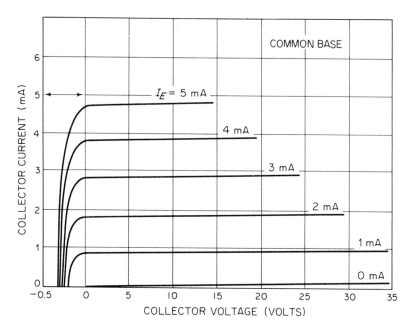

Figure 6-11. Characteristics for the Transistor in Figure 6-10

### 6.4.2 Plotting the CB Load Line

The load line is a straight line on the characteristics graph. This line connects the point of *zero $I_C$ and maximum $V_C$* with the point of *zero $V_C$ and maximum $I_C$*. Using the values in the model circuit, these two points are easily determined. The $V_{CC}$ is $-20$ V. When there is zero $I_C$, there is no voltage drop across $R_C$, and the total $V_{CC}$ appears on the collector. So zero $I_C$ and $-20$ V $V_C$ is the first point. When maximum $I_C$ is present, all of $V_{CC}$ is dropped across $R_C$, and there is zero $V_C$. With the 5 kΩ for $R_C$ the maximum $I_C$ is

$$I_C = \frac{V_{CC}}{R_C}$$
$$= \frac{20 \text{ V}}{5 \text{ k}\Omega}$$
$$= 4 \text{ mA}$$

So the second point is zero $V_C$ and 4 mA $I_C$. When these two points are connected with a straight line, the load line is complete, as illustrated in Figure 6-12.

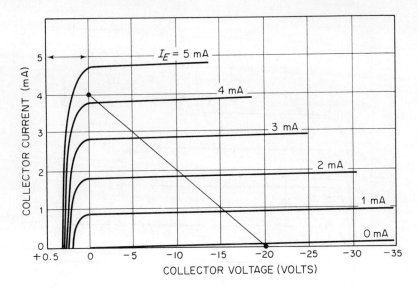

Figure 6-12. Load Line for a CB Amplifier

This load line is a locus of all possible combinations of $V_C$ and $I_C$ with $-20$ V $V_{CC}$ and 5 kΩ $R_C$.

### 6.4.3 Fixing the Operating Point

The operating point describes the quiescent conditions of $V_C$, $I_C$, and $I_E$. Ideally the operating point should be where $V_C$ is half the value of $V_{CC}$. In the circuit under consideration,

$$V_C = \frac{-V_{CC}}{2}$$
$$= \frac{-20 \text{ V}}{2}$$
$$= -10 \text{ V}$$

So the point on the graph where $-10$ V intersects the load line is the ideal operating point. Load line and operating point are illustrated on the graph in Figure 6-13.

Now that the operating point has been selected, the static values of $I_E$ and $I_C$ can be read from the graph. In this case, $I_E$ is about 2.2 mA, and $I_C$ is 2 mA. When the test circuit is built, the bias circuit must be designed to produce these values of current.

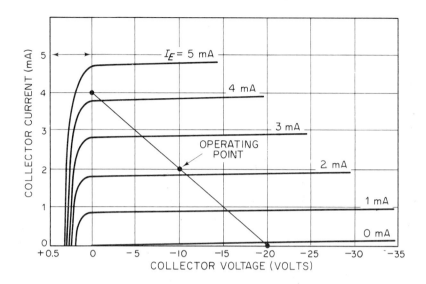

Figure 6-13. Selecting a CB Operating Point

### 6.4.4 Establishing Gain

The incoming signal will cause voltage and current variations to either side of the operating point. The voltage and current value of the input signal is normally known, or else it is easily measured. The current value can be plotted along the load line. The total swing of the input current then reveals the total swing in both $I_C$ and $V_C$.

For explanation purposes, it has been assumed that the input signal has a peak value of 0.2 V and 1 mA. The input current, output voltage, and output current are plotted on the graph of Figure 6-14.

The plot now reveals that a 0.2-V and 2-mA swing of the input signal causes a 9.5-V and 1.9-mA swing of the output. Determining gain is a matter of dividing output by input:

$$A_V = \frac{V_{out}}{V_{in}}$$
$$= \frac{9.5 \text{ V}}{0.2 \text{ V}}$$
$$= 47.5$$

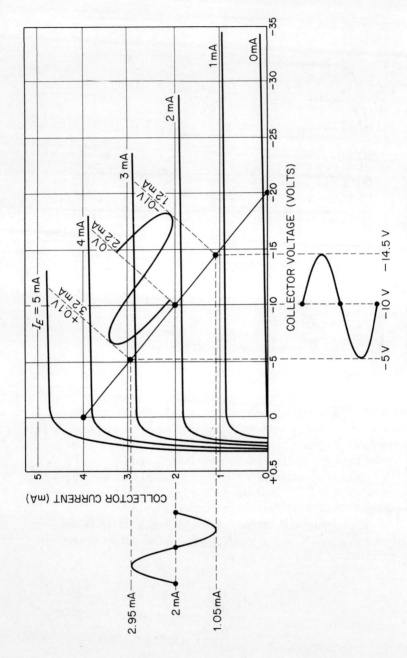

Figure 6-14. Plot of the CB Input and Output

$$A_i = \frac{i_{out}}{i_{in}}$$

$$= \frac{1.9 \text{ mA}}{2 \text{ mA}}$$

$$= 0.95$$

$$G_p = A_v \times A_i$$

### 6.4.5 Common-Emitter Amplifier

Figure 6-15 is a model circuit of a common-emitter amplifier. It has a 24-V supply and a collector load resistor of 4000 Ω. The transistor is an *NPN* that matches the characteristics graph in Figure 6-16.

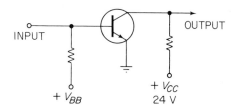

Figure 6-15. Model CE Circuit

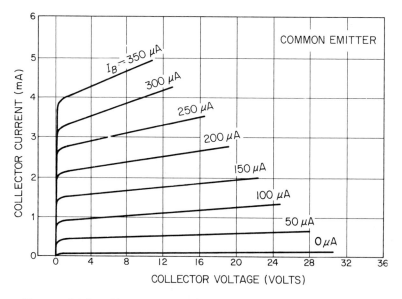

Figure 6-16. Characteristics for the Transistor in Figure 6-15

### 6.4.6 Plotting the CE Load Line

The load line for the common-emitter configuration is plotted exactly as described for the common-base circuit. In this particular case, the load line connects the point of *zero $I_C$ and 24 V $V_C$* to the point of *zero $V_C$ and 6 mA $I_C$*. This load line is plotted in Figure 6-17.

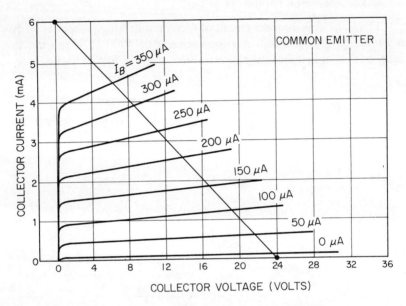

Figure 6-17. Load Line for a CE Amplifier

### 6.4.7 Fixing the Operating Point

The operating point is selected, as before, for the desired quiescent conditions. If it is selected for $V_C$ equal to half of $V_{CC}$, the operating point will be at the intersection of 12 V $V_C$, 3 mA $I_C$, and the load line. This operating point is plotted on the graph of Figure 6-18. Again, when the test circuit is built, it must be biased to place the operating point at this desired position.

### 6.4.8 Establishing Gain

Gain is established by plotting the input and output variations on the load line exactly as previously explained. For this example, let the

# Bipolar Voltage Amplifiers

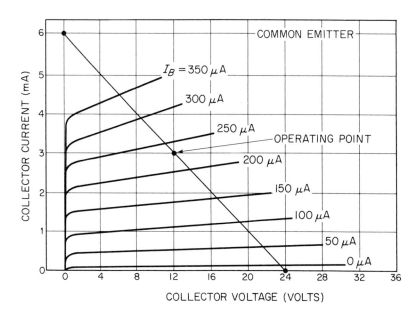

Figure 6-18. Selecting a CE Operating Point

input peak value be 0.3 V and 100 µA. These values of input and the resulting output are plotted on Figure 6-19. The values from the graph are then used to calculate the gain.

$$A_v = \frac{v_{out}}{v_{in}}$$

$$= \frac{10.6 \text{ V}}{0.6 \text{ V}}$$

$$= 17.67$$

$$A_i = \frac{i_{out}}{i_{in}}$$

$$= \frac{2.65 \text{ mA}}{0.2 \text{ mA}}$$

$$= 13.25$$

$$G_p = A_v \times A_i$$

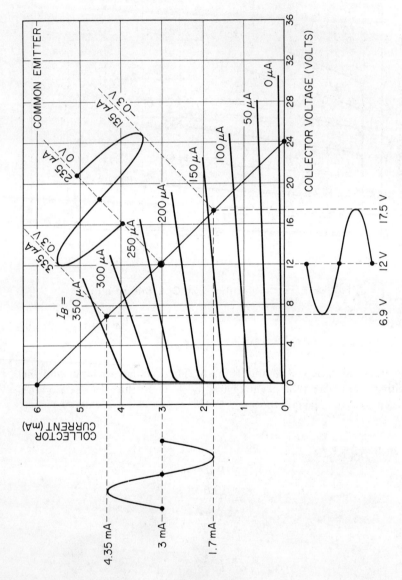

Figure 6-19. Plot of the CE Input and Output

## 6.5 Design Example

The design problem is an audio-frequency voltage amplifier that will meet certain specifications. If the specifications are not listed, designers must make their own specifications from a consideration of the limiting factors they are working under.

### 6.5.1 Specifications

$$Z_i = 20 \text{ k}\Omega$$

$$Z_o = 5 \text{ k}\Omega$$

$$V_{\text{in}} = 0.3\text{-V peak}$$

$$V_{\text{out}} = 3\text{-V peak}$$

$$\text{P.S.} = 12\text{-V battery}$$

$$f = 10 \text{ kHz}$$

$$\text{Fidelity} = \text{maximum}$$

$$\text{Stability} = \text{maximum}$$

### 6.5.2 Model Circuit

After the specifications are analyzed, the type of circuit can be determined, and the model circuit diagram can be drawn. A common-emitter amplifier is nearly always the best choice for an audio-voltage amplifier.

The specifications listed seem to indicate 5000 $\Omega$ each for $R_B$ and $R_C$. For maximum stability the $R_B$ to $R_E$ ratio should be 10. So $R_E$ should be 500 $\Omega$.

The coupling capacitor must pass 50 Hz before its $X_C$ equals half the ohmic value of $R_B$ (assuming low-frequency cutoff to be 50 Hz):

$$\begin{aligned} C &= \frac{1}{2\pi f X_C} \\ &= \frac{0.159}{50 \times 2.5 \times 10^3} \\ &= \frac{159 \times 10^{-6}}{125} \\ &= 1.3 \text{ }\mu\text{F} \end{aligned}$$

A 2-μF capacitor will serve very well; or, if a little more attenuation can be tolerated, a 1-μF capacitor can be used.

The model circuit with the previously determined values is illustrated in Figure 6-20. Selecting a transistor for this circuit is a simple matter. Almost any transistor with a beta more than 10 will suffice.

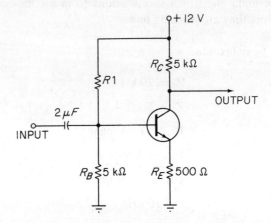

Figure 6-20. Model CE Amplifier

### 6.5.3 Establishing the Operating Point

The high-fidelity requirement indicates that the static $V_C$ should be about half the supply voltage. Using 6 V for $V_C$, the static $I_C$ is

$$I_C = \frac{6 \text{ V}}{5 \text{ k}\Omega}$$

$$= 1.2 \text{ mA}$$

Assuming a beta of 20, the $I_B$ is 0.06 mA. The bias circuit, $R1$, in series with $R_B$, must establish these static conditions.

The value of $I_E$ is $I_C + I_B$:

$$I_E = 1.2 \text{ mA} + 0.06 \text{ mA} = 1.26 \text{ mA}$$

The emitter voltage is 1.26 mA × 500 Ω or 0.63 V.

Assuming a silicon transistor, the value of $V_B$ should be at least 1 V higher than $V_E$. So the base voltage (base to ground) is 1.63 V. This voltage across $R_B$ produces a current of 0.33 mA.

The current through $R1$ is the current through $R_B$ plus the base current.

$$I_{R1} = 0.33 \text{ mA} + 0.06 \text{ mA}$$
$$= 0.39 \text{ mA}$$

The voltage across $R1$ is the supply voltage less the voltage across $R_B$.

$$V_{R1} = 12 \text{ V} - 1.63 \text{ V}$$
$$= 10.37 \text{ V}$$

This voltage divided by the current through $R1$ indicates the ohmic value of $R1$:

$$R1 = \frac{10.37 \text{ V}}{0.39 \text{ mA}}$$
$$= 26.6 \text{ k}\Omega$$

This resistor value can be rounded off to the nearest standard, which is 27 k$\Omega$. When this value is entered on the model circuit it becomes that in Figure 6-21.

### 6.5.4 Analysis and Modification

The model circuit should be analyzed for input impedance, output impedance, voltage gain, and current gain. If this analysis reveals any

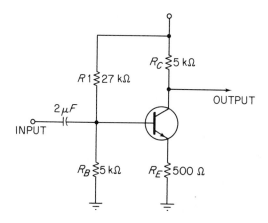

Figure 6-21. Model Circuit with Bias Resistors

of these items to be out of tolerance, modification of the proper circuit component must be made.

### 6.5.5 Test Circuit

After the model circuit has been analyzed and the necessary modifications made, it can be used as a diagram to construct the test circuit.

The best way to build the test circuit is to solder each component in place. If soldering is not convenient, clamps can be used. In either case, each joint must have a good electrical connection.

## 6.6 Testing and Troubleshooting

Building the test circuit is not the final design step. Now the test circuit must be checked to see if it performs according to the calculated results. If it does not meet expectations, the trouble must be located and repaired.

### 6.6.1 Testing Static Values

The test circuit is now complete. It should be the same as the final model circuit shown in Figure 6-21. The first testing step is to check static values of direct voltages, and from these measurements to determine if current values and operating point are within tolerance. A $\pm 10$ percent tolerance on all values is very good. A tolerance of $\pm 20$ percent is generally acceptable. Over 20 percent deviation requires circuit changes. These measurements can be made with a multimeter. Do not attempt to obtain perfect readings. Remember that the meter also has a generous tolerance.

### 6.6.2 Testing for Distortion and Gain

Now the signal generator is connected to the input of the test circuit. The generator must be set for the signal amplitude output that the test circuit was designed to receive. *This is important:* All signal generators change signal amplitude as the frequency is changed. With each change of frequency, check the output signal amplitude and readjust it as indicated. To start, set the signal generator frequency to the center frequency of the design circuit. In the example used here, that is 10 kHz.

The oscilloscope is then connected to the input for observing and measuring the input signal. This is a voltage amplifier, so voltages are of primary concern. Observe the exact waveshape and measure the peak-to-peak amplitude.

Move the oscilloscope leads to the output terminals, and repeat the procedure for the output signal. Any deviation of waveshape between input and output (except amplitude and phase inversion) is distortion. If distortion is obvious, it indicates a circuit problem.

Determining voltage gain is simply a matter of dividing output voltage by input voltage, and peak-to-peak values directly from the oscilloscope are as good as any.

If a dual-trace oscilloscope is available, one channel should be connected to the input and the other to the output. This is the ideal way to compare two signals.

### 6.6.3 Testing for Frequency Response

Assuming that the center frequency passed the test for distortion and gain, the circuit should then be checked for total frequency response.

Lower the frequency of the signal generator down to the low-frequency cutoff point. For audio circuits, this is about 50 Hz. Reset the output signal level. Measure and record frequency and the output signal amplitude (peak to peak is fine). Increase the frequency in steps of about 100 Hz; reset amplitude, and measure and record frequency and output amplitude after each step. When a constant amplitude is detected in the output, change frequency in larger steps until amplitude starts to decrease again. Continue in smaller steps until output drops to about half the center value. When all these readings are plotted on a graph of amplitude versus frequency, they reveal the total frequency response of the test circuit from low-frequency cutoff to high-frequency cutoff.

### 6.6.4 Troubleshooting

Any test result that is out of tolerance indicates a circuit defect that must be corrected. It may be a malfunction or simply the wrong value for a component. Circuit failure usually reveals itself in one of three ways: no output, reduced output, or distorted output.

*No output* is caused from failure of the transistor or a failure in the bias circuits. A recheck of direct voltage will probably pinpoint the defective component. In test circuits, the trouble is very likely a loose connection. Make sure that all connections are secure before exchanging components. Sometimes a resistor, capacitor, or even a transistor is destroyed in the process of soldering. These are next to be checked if all connections are proper.

A *reduced output* generally indicates a bad transistor, but it may be caused by improper bias. If bias is correct, check the transistor. An in-circuit check should be made if possible; if not, at least a simulated operational check should be made. If the transistor is good, the load resistor is probably bad—not open or shorted, just out of tolerance.

A *distorted output* usually indicates too much gain. Too much gain can be caused by any one of several faults. An open biasing resistor, faulty decoupling capacitors, and too much regenerative feedback are among the most likely. The transistor may have changed its amplifying characteristics, but this is less likely.

The general procedures for locating a trouble are:

1. Perform a close visual inspection.
2. Check direct voltages.
3. Measure resistances.
4. Check capacitors (an ohmmeter check is sufficient).
5. Check the transistor (dynamic test).
6. Check for secondary damage.

After repair, repeat the complete testing process.

CHAPTER SEVEN

# Unipolar Voltage Amplifiers

All unijunction transistors have become popular voltage-amplifying devices. This grouping includes both JFETs and IGFETs. This chapter examines some of the design features of using unipolar transistors as audio-voltage amplifiers. Methods of analysis, testing, and troubleshooting are included along with design examples.

## 7.1 Basic JFET Amplifier

The basic JFET audio-voltage amplifier is very similar to the biasing circuits discussed in Chapter Three. In some special cases a common gate circuit can be used as a voltage amplifier, but in many cases the input impedance for a common gate is too small for practical applications. The common drain configuration cannot be used because it does not amplify voltage. The voltage gain of a common drain circuit is always less than unity. So the majority of JFET voltage amplifiers are in the common-source configuration.

### 7.1.1 Model JFET Circuit

Since the common source is the most used circuit, the example analyzed here is a common-source amplifier. A model circuit of a simple JFET audio-voltage amplifier is shown in Figure 7-1 for purpose of analysis. The amplifying device in this circuit is an *N*-channel JFET. Otherwise, the circuit is little different from that of the bipolar common-emitter amplifier. The primary design differences are in the characteristics of the transistor and in the value of the circuit components.

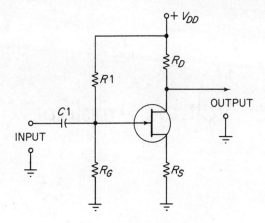

Figure 7-1. Model JFET Circuit

The transistor has a combination of fixed bias and self-bias. The fixed bias is furnished by the voltage divider, consisting of $R1$ and $R_G$. The self-bias is furnished by $R_S$, which also provides degenerative feedback to improve the fidelity. The input is $RC$-coupled to the gate by $C1$ and $R_G$, and the output is developed across $R_D$.

The supply voltage, $V_{DD}$, is a positive potential that causes all elements of the transistor to be positive with respect to ground. The gate will draw current if it is allowed to go positive with respect to the source. This gate current would cause distortion, so the gate is kept at a lower potential than the source. The bias is the difference between the voltage drop across $R_S$ and the voltage drop across $R_G$. The relative values of $R1$ and $R_G$ are chosen to produce the required value of static drain current.

### 7.1.2 Estimating Component Values

As with the bipolar circuit, the next step is to estimate the values of the various components. These estimations should be more than a rough guess; they should be based on physical limitations and circuit specifications.

Since this is an audio-voltage amplifier, high fidelity is undoubtedly one of the specifications. Matching input and output impedance as closely as is reasonably possible is another automatic consideration. The amount of gain is a probable specification, and the available power supply voltage may be a limiting factor. For this example, assume a regulated 30-V power supply.

The input resistance to a JFET in the common-source configuration is so high that it need not enter into the calculations. The value of $R_G$ then becomes the circuit input impedance. Suppose that the output of the previous stage is 5000 Ω. Then this means that $R_G$ for this stage should be 5000 Ω. The value of $R1$ is not normally considered because its value is usually much larger than that of $R_G$. However, if $R1$ is less than 10 times the value of $R_G$, then the input impedance is $R1$ in parallel with $R_G$.

The value of $C1$ is selected for the low-frequency cutoff point. Remember that the low-frequency cutoff occurs at the frequency which causes $X_C$ of $C1$ to be half the value of $R_G$. Using 50 Hz of the low-frequency cutoff, the value of $C1$ is

$$C1 = \frac{1}{2\pi f X_C}$$
$$= \frac{0.159}{50 \times 2.5 \times 10^3}$$
$$= \frac{159 \times 10^{-6}}{125}$$
$$= 1.3 \ \mu F$$

A 1-μF capacitor is satisfactory but a 2-μF capacitor is better.

The output impedance should match the input impedance of the next stage. Suppose that the input impedance to the next stage is 3000 Ω. Then the trial value of $R_D$ can be set to 3000 Ω. Remember that the output impedance of the circuit is $R_D$, in parallel with $r_d$ (where $r_d$ is the internal resistance from drain to source). When the value of $r_d$ is determined, the value of $R_D$ might have to change to a higher or a lower value. The value of $r_d$ can range from 20 kΩ to more than 100 kΩ.

In selecting a value for $R_S$ there must be a compromise between gain and stability. The gain is approximately equal to $R_D$ divided by $R_S$, while stability depends on the ratio of $R_G$ to $R_S$. With $R_G$ of 20 times $R_S$, there is optimum gain. With $R_G$ 10 times $R_S$, there is optimum stability. A good rule is to keep $R_S$ as close as possible to one tenth of both $R_G$ and $R_D$. In this case, $R_D$ is 3000 Ω and $R_G$ is 5000 Ω. The average of these two is 4000 Ω. Using this average, one tenth of that is 400 Ω, so $R_S$ can be estimated as 400 Ω.

Assuming that class A operation with minimum distortion is a desirable feature, the operating point should set the static $V_D$ to half the value of $V_{DD}$. In this case, $V_{DD}$ is 30 V, and the bias must establish a static $I_D$ to drop half this value (15 V) across $R_D$. Static $I_D$ becomes

$$I_D = \frac{15 \text{ V}}{3 \text{ k}\Omega}$$

$$= 5 \text{ mA}$$

This 5 mA through the 3-k$\Omega$ drain load resistor, $R_D$, will drop 15 V across $R_D$ and leave 15 V on the drain.

The same 5 mA produces a voltage drop across $R_S$ that is

$$V_{R_S} = 5 \text{ mA} \times 400 \text{ }\Omega$$

$$= 2 \text{ V}$$

The input bias is $V_{R_G} - V_{R_S}$, and this must be a negative value for all values of the input signal. Furthermore, the peaks of the input signal must not be allowed to drive the transistor below the pinch-off region. Using an ac signal of 1-V peak as an input, the static voltage drop across $R_G$ can be estimated as +1 V.

The value of $R1$ then must be such that it will place a 1-V drop across $R_G$. The ratio of voltage is $R_G = 1$ V and $R1 = 29$ V; so $R1$ must be 29 times the value of $R_G$. The value of $R_G$ has been set at 5000 $\Omega$, so

$$R1 = 29 \times 5000 \text{ }\Omega$$

$$= 145 \text{ k}\Omega$$

The transistor selected is an $N$-channel JFET that has a $Y_{fs}$ of 4000 $\mu$mhos and a $Y_{os}$ of 10 $\mu$mhos. When all these estimated values are recorded on the model circuit of Figure 7-1, the results are as illustrated in Figure 7-2.

## 7.2  JFET Equivalent-Circuit Analysis

The estimated values on the model circuit provide information for a theoretical analysis. The analysis should reveal any needed changes on the model circuit. The first analysis is through the use of an equivalent circuit.

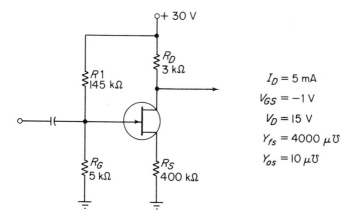

Figure 7-2.  Model JFET Circuit with Trial Components

### 7.2.1  Constructing the Circuit

The JFET equivalent circuit is simpler than the bipolar equivalent circuit. At high frequencies the capacitance becomes a factor, but for audio frequencies only resistances and voltages need be considered. The low-frequency equivalent circuit for the diagram in Figure 7-2 is illustrated in Figure 7-3.

Internally, the JFET consists of $R_{GS}$, a voltage generator, and $r_d$. Without feedback from $R_S$, the circuit would be $R_G$ in parallel with $R_{GS}$ and $r_d$ in parallel with $R_D$. Notice that there is no voltage generator in the equivalent circuit. The output voltage is the change in $I_D$ multiplied by $R_D$. And $I_D$ at any instant is $g_m$ times $V_{in}$.

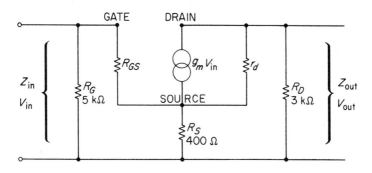

Figure 7-3.  Low-Frequency Equivalent of Figure 7-2

### 7.2.2 Input Impedance

The true input impedance of this circuit is $R_G$ in parallel with $R_{GS} + R_S$. But $R_{GS}$ is such a large resistance that the combination of $R_{GS} + R_S$ has very little effect on the input impedance. So the circuit input impedance is very nearly the same as the value of $R_G$.

$$Z_{in} = R_G$$

Since the value of $R_G$ has been selected to match the output impedance of the previous stage, there is nothing in this equivalent circuit to indicate any change.

### 7.2.3 Output Impedance

The true output impedance is $R_D$, in parallel with $r_d + R_S$. The value of $r_d$ may range from 20 k$\Omega$ to 100 k$\Omega$. The manufacturer's data sheet may *not* list the value of $r_d$; it may list $Y_{os}$ instead, as shown in Figure 7-2. This $Y_{os}$ is the output admittance, and it is the reciprocal of $r_d$. In this case, $Y_{os}$ is 10 $\mu$mhos, so

$$r_d = \frac{1}{Y_{os}}$$
$$= \frac{1}{10 \times 10^{-6}}$$
$$= \frac{1 \times 10^6}{10}$$
$$= 100 \text{ k}\Omega$$

The 400 ohms of $R_S$ in series with $r_d$ is of no significance. The 100 k$\Omega$ of $r_d$ is so large in comparison to $R_D$ of 3 k$\Omega$ that the output impedance for all practical design considerations is equal to $R_D$:

$$Z_{out} = R_D$$

### 7.2.4 Simplified Equivalent Circuit

The foregoing analysis concerning $R_G$ and $R_D$ will be true in nearly all audio-voltage amplifier designs. The value of $R_{GS}$ is so large that it is never considered. The value of $r_d$ becomes significant only when it is less than 10 times the value of $R_D$. In most cases, $r_d$ can be ignored.

The equivalent circuit of Figure 7-3, and of most JFET audio-voltage amplifiers, can then be simplified, as illustrated in Figure 7-4.

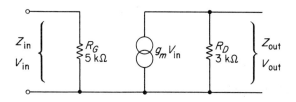

Figure 7-4. Simplified Equivalent JFET Circuit

### 7.2.5 Voltage Gain

The voltage gain is simply output voltage divided by input voltage. The input voltage, $V_\text{in}$, is the signal voltage from the previous stage. In this case, an input of 1 V peak was specified.

The output voltage is the voltage across $R_D$. Some authors place a minus sign before the output voltage to indicate phase inversion. It is not practiced in this book because the reader needs no reminder that a common collector and a common source always have voltage inversion. Without feedback this output voltage is $\Delta I_d \times R_D$ and $I_d$ is $g_m \times V_\text{in}$. The parameter $g_m$ is transconductance, but it is not always listed as $g_m$. On some data sheets $Y_{fs}$ is specified. The forward transfer admittance is $Y_{fs}$ and this means the same as transconductance.

$$g_m = Y_{fs}$$

With feedback, part of this voltage is dropped across $R_S$. The voltage-gain equation is

$$A_v = \frac{g_m R_D}{(g_m + 1) R_S}$$

For practical purposes, this equation is shortened to

$$A_v = \frac{R_D}{R_S}$$

In this case,

$$A_v = \frac{R_D}{R_S}$$

$$= \frac{3000}{400}$$

$$= 7.5$$

#### 7.2.6 Summary of the Equivalent-Circuit Analysis

It should be concluded that, in most low-frequency voltage circuits, $R_G$ is the input impedance and $R_D$ is the output impedance. These values can be selected directly to match the impedance of the previous stage and the following stage. These values can then be rounded off to the nearest standard resistor value.

The voltage gain is simply

$$A_v = \frac{V_\text{out}}{V_\text{in}}$$

With a source resistor for feedback, voltage gain is approximately

$$A_v = \frac{R_D}{R_S}$$

Above all, the analysis should have emphasized the simplicity of this type of circuit.

### 7.3  JFET Load-Line Analysis

The equivalent circuit analysis turns out almost too simple for elaboration. The same JFET circuit (Figure 7-2) is now analyzed with a load line.

#### 7.3.1 Constructing the Load Line

The first required item is a family of characteristics curves for this particular JFET. The graph in Figure 7-5 represents the transistor in Figure 7-2.

The load line is constructed exactly as described in Chapter Six for the bipolar transistor. It connects the point of minimum $I_D$ and maximum $V_D$ to the point of maximum $I_D$ and minimum $V_D$.

# Unipolar Voltage Amplifiers

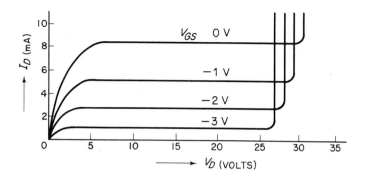

Figure 7-5.  JFET Characteristics

Maximum $V_D$ is 30 V and maximum $I_D$ is

$$I_D = \frac{V_{DD}}{R_D}$$
$$= \frac{30 \text{ V}}{3 \text{ k}\Omega}$$
$$= 10 \text{ mA}$$

This load line and the operating point are indicated on the graph of Figure 7-6. The operating point was previously described as $I_D = 5$ mA, $V_D = 15$ V, and $V_{GS} = -1$ V.

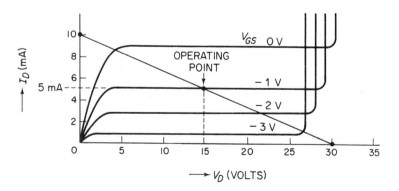

Figure 7-6.  JFET Load Line

### 7.3.2 Establishing Gain

It is a simple matter to plot input and output voltages and determine the gain of the circuit. The voltage waveshapes are plotted on the graph in Figure 7-7. As the input swings 2 V peak to peak, the output swings 18 V peak to peak. According to the graph, the voltage gain is

$$A_v = \frac{V_{out}}{V_{in}}$$
$$= \frac{18 \text{ V}}{2 \text{ V}}$$
$$= 9$$

This is somewhat greater than the 7.5 obtained in the equivalent-circuit analysis. However, with the load line, the feedback was disregarded. The actual gain is a little less than the 9 indicated here.

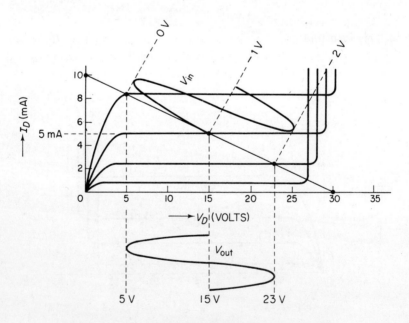

Figure 7-7. Plot of the JFET Input and Output

## 7.4 Basic MOSFET Amplifier

The $N$-channel enhancement MOSFET has been selected as the sample audio-frequency voltage amplifier using an insulated gate transistor. This is a very popular transistor for voltage amplification and the operation is very similar to that of the JFET.

### 7.4.1 Model MOSFET Circuit

The model circuit for a MOSFET audio-frequency voltage amplifier is little different from the other model circuits already discussed. The difference is in the amplifying device and some of the component values. The circuit is the common source since this is the usual voltage amplifier configuration. The model circuit without component values is illustrated in Figure 7-8. Since this is an $N$-channel enhancement MOSFET, both drain and gate must be positive with respect to the source. For all practical purposes there is no gate current, since the input resistance is on the order of $10^{15}$ Ω. Under these circumstances, pure self-bias is impossible, even if it were desirable. The external fixed bias is supplied by the bleeder network, consisting of $R1$ and $R_G$. All audio-voltage amplifiers should have reasonably high fidelity, so $R_S$ is included for degenerative feedback.

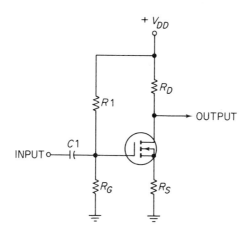

Figure 7-8. Model MOSFET Circuit

### 7.4.2 Estimating Component Values

The voltage source is a regulated 30 V, and this fixes the value of $V_{DD}$. The output impedance of the previous stage is 47 kΩ and the input impedance of the next stage is 4700 Ω. This provides trial values for $R_G$ and $R_D$. The input signal is a sine wave of 1 V peak amplitude and the output must be at least 3 V peak. This is a low amplification requirement, so a rather heavy degenerative feedback should be possible for maximum fidelity.

The maximum size of $R_S$ can be calculated by using the approximate gain formula:

$$A_v = \frac{R_D}{R_S}$$

$$R_S = \frac{R_D}{A_v}$$

In this case, a gain of 3 is minimum, so

$$R_S = \frac{4700}{3}$$

$$= 1566 \ \Omega$$

Considerable rounding off has been done here, so the trial value of $R_S$ should be this calculated value reduced by about one third. So 1000 Ω is a good estimate.

The transistor has the characteristics indicated in Figure 7-9. The operating point should produce a $V_D$ about half the value of $V_{DD}$. From the chart this operating point can be estimated as $V_D = 15$ V and $V_{GS} = 2$ V. These conditions produce an $I_D$ of 3 mA. The 3 mA of $I_D$ through $R_S$ of 1000 Ω produces a voltage drop of 3 V. The voltage across $R_G$ must be this 3 V across $R_S$ plus the 2 V of $V_{GS}$. So $R_G$ must drop the remaining 25 V of $V_{DD}$.

The current through $R_G$ and through $R1$ is

$$I = \frac{E}{R}$$

$$= \frac{5 \text{ V}}{47 \text{ k}\Omega}$$

$$= 0.1 \text{ mA}$$

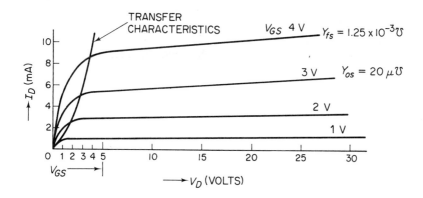

Figure 7-9. Characteristics of an N-channel Enhancement MOSFET

The resistance of $R1$ is

$$R1 = \frac{E}{I}$$
$$= \frac{25 \text{ V}}{0.1 \text{ mA}}$$
$$= 250 \text{ k}\Omega$$

The closest standard resistance is 270 k$\Omega$, and that should be plenty close.

The coupling capacitor should have a low-frequency cutoff at about 50 Hz. With an $R_G$ of 47 k$\Omega$ a capacitor that offers 23.5 k$\Omega$ (half of 47 k$\Omega$) to 50 Hz will be the proper capacitance.

$$C1 = \frac{1}{2\pi f X_C}$$
$$= \frac{0.159}{50 \times 23.5 \times 10^3}$$
$$= \frac{159 \times 10^{-6}}{1175}$$
$$= 0.14 \times 10^{-6}$$
$$= 0.14 \text{ }\mu\text{F}$$

The next larger standard value capacitor will serve fine. So let the coupling capacitor be 0.2 μF.

Now that the value of all the components has been considered, they can be entered on the diagram of the model circuit as illustrated in Figure 7-10.

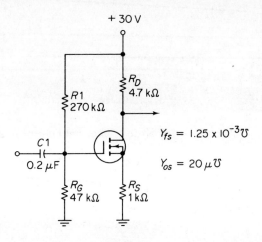

Figure 7-10. Model MOSFET Circuit with Trial Components

## 7.5 MOSFET Equivalent-Circuit Analysis

The MOSFET equivalent circuit is the same as that previously used for the JFET amplifier. The only difference is the value of the resistors and the characteristics of the transistor. The equivalent circuit for Figure 7-10 is illustrated in Figure 7-11.

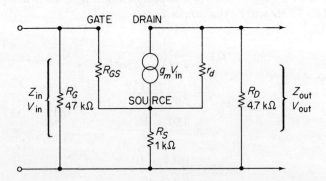

Figure 7-11. Low-Frequency Equivalent of Figure 7-10

## 7.5.1 Impedance

The value of the internal input impedance, $R_{GS}$, is very nearly an open circuit. Only the circuit value of $R_G$ and $R1$ need to be considered for the input impedance. Since $R1$ is less than 10 times the value of $R_G$, the input signal has an impedance of $R1$, in parallel with $R_G$.

$$Z_{in} = \frac{R1 R_G}{R1 + R_G}$$

$$= \frac{(270 \times 10^3)(47 \times 10^3)}{270 \times 10^3 + 47 \times 10^3}$$

$$= \frac{12{,}690 \times 10^3}{317}$$

$$= 40 \text{ k}\Omega$$

This is less than the desired 47 k$\Omega$ for a perfect match, but the mismatch is within the 25 percent tolerance that is normally allowed. So let $R_G$ stand at 47 k$\Omega$.

The value of $r_d$ is the reciprocal of $Y_{os}$, and in this case, $Y_{os}$ is 20 $\mu$mhos.

$$r_d = \frac{1}{20 \times 10^{-6}}$$

$$= 50 \text{ k}\Omega$$

Since this $r_d$ is more than 10 times the value of $R_D$, only $R_D$ needs to be considered for the output impedance. Then, for most practical applications, the simplified equivalent circuit as represented in Figure 7-12 is accurate enough.

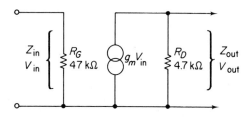

Figure 7-12. Simplified Equivalent MOSFET Circuit

### 7.5.2 Voltage Gain

The voltage gain is roughly

$$A_v = \frac{R_D}{R_S}$$

$$= \frac{4.7 \text{ k}\Omega}{1 \text{ k}\Omega}$$

$$= 4.7$$

More precisely,

$$A_v = \frac{g_m R_D}{(g_m + 1)R_S}$$

The $Y_{fs}$ for this transistor was given as $1.25 \times 10^{-3}$. Either method seems to indicate a voltage gain in excess of the necessary 3 that is required.

## 7.6 MOSFET Load-Line Analysis

Using the characteristics graph in Figure 7-9 and the information from the circuit in Figure 7-10, a load line is easily constructed.

### 7.6.1 Constructing the Load Line

The maximum $V_D$ is 30 V, which is plotted on the zero-$I_D$ line. The maximum $I_D$ is

$$I_D = \frac{V_{DD}}{R_D}$$

$$= \frac{30 \text{ V}}{4.7 \text{ k}\Omega}$$

$$= 6.4 \text{ mA}$$

This 6.4 mA is plotted on the zero-$V_D$ line. When these two points are connected with a straight line, the result is as illustrated in Figure 7-13. Notice that the operating point is not precisely where previous analysis indicated. Keeping the $V_{GS}$ and $I_D$ at previous values, the operating point is the intersection of $V_{GS} = 2$ V and the load line. This also indicates the selected 3 mA of $I_D$. But the $V_D$ is 16 V instead of 15 V. This small difference is of no concern.

# Unipolar Voltage Amplifiers

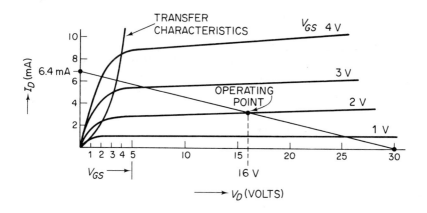

Figure 7-13.  MOSFET Load Line

### 7.6.2  Plotting the Input and Output

The input voltage was specified as 1 V peak. This voltage can now be plotted about the operating point on the load line. The output voltage is then plotted. Input and output plots are indicated on the graph of Figure 7-14.

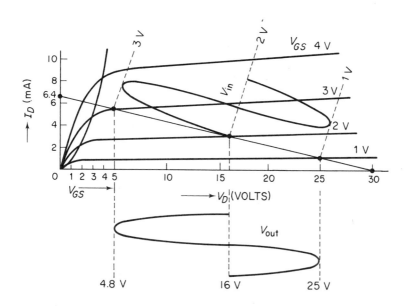

Figure 7-14.  Plot of the MOSFET Input and Output

According to this plot, a 2-V peak-to-peak input causes the output to swing from 4.8 V to 25 V for a peak-to-peak amplitude of 20.2 V. This appears to be a gain of 10.1, but the results are misleading. This circuit has a strong degenerative feedback, which reduces the gain at least 50 percent. The feedback brings the gain down to about 5, and that is in the area of the results obtained previously.

## 7.7 Test Circuits

The limited analysis and approximations applied to the model circuits for both JFET and MOSFET indicate that they will perform according to specifications. The next step is to build the circuits and test their performance.

## 7.8 Design Example

The design problem is an audio-frequency voltage amplifier that must meet the following specifications.

### 7.8.1 Specifications

$$Z_i = 4 \text{ k}\Omega$$

$$Z_o = 3 \text{ k}\Omega$$

$$V_{\text{in}} = 1 \text{ V peak}$$

$$V_{\text{out}} = 4 \text{ V peak (min)}$$

$$\text{P.S.} = 24 \text{ V regulated}$$

$$f = 5 \text{ kHz}$$

$$\text{Fidelity} = \text{optimum}$$

$$\text{Stability} = \text{optimum}$$

### 7.8.2 Model Circuit

Either a JFET or a MOSFET will serve this circuit. The selected transistor is an $N$-channel JFET with the following characteristics:

$$Y_{os} = 10 \text{ }\mu\text{mhos}$$

$$Y_{fs} = 3000 \text{ }\mu\text{mhos}$$

# Unipolar Voltage Amplifiers

The 4 kΩ of input impedance indicates an $R_G$ on the order of 4 kΩ. The output impedance is fixed at 3 kΩ; so $R_D$ will be 3 kΩ. Optimum fidelity and stability indicated some degenerative feedback. The circuit should be a common source configuration with $R_S$ about one tenth of $R_D$. So $R_S$ is 300 Ω. The initial circuit is illustrated in Figure 7-15.

The coupling capacitor should pass frequencies as low as 50 Hz before its $X_C$ reaches half the value of $R_G$. In this case, half of $R_G$ is 2 kΩ.

$$C = \frac{1}{2\pi f X_C}$$

$$= \frac{0.159}{50 \times 2 \times 10^3}$$

$$= \frac{159 \times 10^{-6}}{100}$$

$$= 1.59 \; \mu F$$

A smaller value capacitor would cause attenuation at the low frequencies, but any capacitor of 2 μF or larger is satisfactory.

### 7.8.3 Establishing the Operating Point

The static $V_D$ should be about 12 V for this circuit. This requires an $I_D$ of 4 mA. This 4 mA will drop 12 V across $R_D$ and 1.2 V across $R_S$.

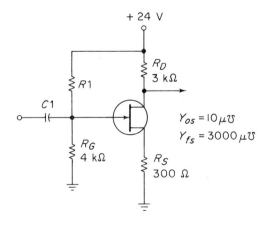

Figure 7-15.  Initial JFET Circuit

Assuming that a $V_{GS}$ of $-1$ V will provide the required 4 mA of $I_D$, the value of $R1$ can now be calculated. The voltage across $R_G$ is 0.2 V, making $V_G$ 1 volt negative with respect to $V_S$. The voltage drop across $R1$ must be about 24 times as large as that across $R_G$. The value of $R1$ then needs to be 24 times the value of $R_G$. Since $R_G$ is 4 kΩ, $R1$ should be 96 kΩ. The remaining estimated values can be entered on the diagram as indicated in Figure 7-16.

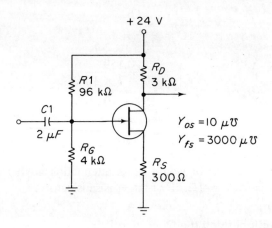

Figure 7-16.  Initial JFET Circuit with Test Values

### 7.8.4   Test Circuit

The diagram in Figure 7-16 can be used to construct the test circuit.

## 7.9   Testing and Troubleshooting

The output of the test circuit should be connected to a 3-kΩ load, and the input to a signal generator. The signal generator output impedance should be set to 4 kΩ. It would be helpful if $R1$ and $R_S$ can be decade or variable resistors.

### 7.9.1   Testing Static Values

Without an input, use the multimeter to measure the static voltages, and vary $R1$ and $R_S$ until all direct-voltage values are reasonably close to the predetermined values.

### 7.9.2 Testing for Distortion and Gain

Now set the signal generator output to 1 V peak and 4 kΩ at a frequency of 5 kHz. Measure the peak-to-peak swing across $R_G$ and across $R_D$. This may be done with a multimeter, but the oscilloscope is better. At any rate, an oscilloscope is necessary for observing waveshapes and checking for distortion. This is the center frequency, and there should be ample gain without any noticeable distortion. Compare input and output for a faithful reproduction of the waveshape.

### 7.9.3 Testing for Frequency Response

Set the signal generator to 50 Hz, and reset both amplitude and output impedance. Change frequency in reasonable steps and check for gain and distortion over a band from 50 Hz to about 12 kHz. Plot a graph of responses as described in Chapter Six.

### 7.9.4 Troubleshooting

*No output* means that the transistor has failed or one of the resistors is open. The most likely trouble is a bad connection. A recheck of the static direct voltages will probably isolate the trouble. This trouble could be an open coupling capacitor, but this is not likely. An open lead to the capacitor would produce the appearance of an open capacitor.

*Reduced output* comes from a faulty transistor. Check the static voltages. If voltages are correct, check the transistor. Use an in-circuit check when possible. If the in-circuit check cannot be made, use a dynamic check under simulated circuit conditions.

*Distorted output* is an indication of too much gain. Check for proper biasing. If static voltages check out within tolerance, check the transistor. If the transistor is good, increase the value of $R_S$ and readjust the operating point.

General procedures should be followed as detailed in Chapter Six for locating trouble. And after repair, repeat the complete testing process.

CHAPTER EIGHT

# Power Amplifiers

The final stage in most audio-frequency circuits is some type of speaker. The audio-frequency signal must pass through a stage of power amplification to attain the power required to drive the speaker. The AF *power amplifier* is a low-power circuit when compared to the high-power circuits used in transmitters, but the design considerations are common in many ways. The operating limits of the transistor have been largely ignored in AF voltage amplifiers because there is little chance of exceeding any of the maximum specifications.

The maximum ratings of a transistor are prominent factors in power amplifiers. These ratings indicate maximum values of current, voltage, temperature, and power. The designer of a power-amplifier stage must be fully cognizant of the permissible confines of transistor operation.

This discussion is centered on bipolar transistors as power amplifiers because most present-day power amplifiers use bipolar transistors. However, field-effect transistors are finding their way into the power-amplification area. Much of this discussion applies equally well to FET power amplifiers.

## 8.1 Power Limitations

Power is the product of voltage times current. The higher the power, the more the heat that must be dissipated. The power limitations of a transistor are determined by several factors, including size of the elements, type of case, size of the case, and how the case is mounted. The designer must realize these limitations and work within them to avoid circuit failure.

### 8.1.1 Power Ratings

The transistor data sheet specifies absolute maximum ratings at a case temperature of 25°C. These ratings include collector–base voltage, collector–emitter voltage, emitter–base voltage, continuous collector current, continuous base current, continuous power dissipation, operating case temperature, and a derating factor for higher temperatures. Sample ratings for a power transistor follow:

Collector–base voltage: 100 V.
Collector–emitter voltage: 70 V.
Emitter–base voltage: 7 V.
Continuous collector current: 15 A.
Continuous base current: 7 A.
Continuous power: 115 W.
Operating range: −65° to 200°C.

*Note:* Derate 0.6 W/°C for case temperatures in excess of 25°C.

It should be remembered that the case temperature is generally higher than the ambient air temperature. This is discussed in detail in Section 5.7. So when case temperature is specified, some calculations may be required to relate this to the ambient air temperature.

### 8.1.2 Maximum Power Versus Case Temperature

The higher the case temperature, the lower the maximum power. This fact is clearly indicated by the derating factor. In addition to the derating factor, most manufacturers publish a graph that indicates the power and temperature limitations. Such a graph is shown in Figure 8-1.

This sample graph indicates that this transistor can handle 100 W of power when the case temperature is 25°C or less. As the case temperature increases, the maximum power decreases. At a case temperature of 115°C, the maximum power is reduced by half. At a case temperature of 200°C, any power at all is likely to destroy the transistor.

### 8.1.3 Plotting the Maximum Power Curve

The maximum power of the transistor is determined after considering the operating case temperature. This maximum power curve should then be plotted on the static characteristics graph for that transistor. The maximum power curve is a locus of all values of $V_C$ and $I_C$ that

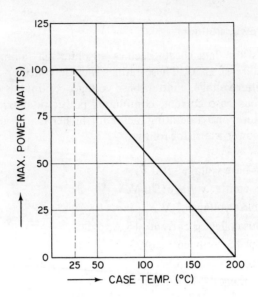

Figure 8-1. Power and Temperature Limitations

will produce the maximum permissible power. A sample of such a curve is illustrated in Figure 8-2. The transistor represented by this graph is capable of handling 100 W at a case temperature of 25°C. The maximum power curve is plotted for this 25°C case temperature. If the operating case temperature is higher than 25°C, the derating factor should be applied before constructing the curve. After the derating factor has been applied, the curve is located by plotting several points where $I_C \times V_C$ = maximum power.

The easiest way to locate the points is to assume several values of $V_C$ and calculate the $I_C$. In this case, maximum power is 100 W, so

$$I_C = \frac{100 \text{ W}}{V_C}$$

The points plotted are:

$V_C = 80$ V  $\qquad I_C = 1.25$ A
$V_C = 50$ V  $\qquad I_C = 2$ A
$V_C = 40$ V  $\qquad I_C = 2.5$ A
$V_C = 30$ V  $\qquad I_C = 3.3$ A

Power Amplifiers

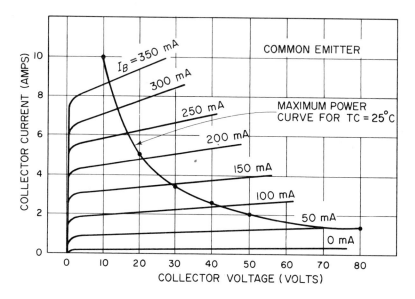

Figure 8-2. Maximum Power Curve

$$V_C = 20 \text{ V} \quad I_C = 5 \text{ A}$$
$$V_C = 10 \text{ V} \quad I_C = 10 \text{ A}$$

The curve is completed by connecting the plotted points with a smooth curve.

This maximum power curve outlines the power limits, which must never be exceeded for a significant period of time.

### 8.1.4 Load Line

The power amplifier load line should be plotted on the same graph that the maximum power curve is on. The load line may touch, but must never cross, the power curve. The proper relationships of the load line and the power curve is illustrated in Figure 8-3. This load line runs adjacent to the hyperbola of the power curve. The two lines touch only briefly at one point; they do *not* cross. A good operating point for this circuit is the point where 50 V $V_C$ intersects 4 A $I_C$. Notice that this is also the point where the load line touches the power curve. This load line represents the maximum circuit conditions for this transistor. Any increase of either $V_C$ or $I_C$ will cause the load line to cross the curve.

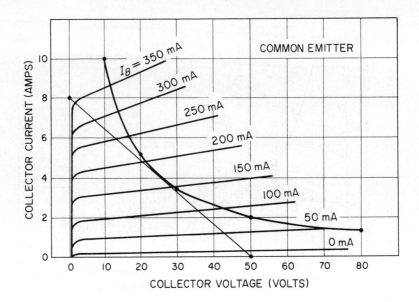

Figure 8-3. Relationship of a Load Line and a Power Curve

## 8.2 Class A Amplifier with a Resistive Load

Nearly all voltage amplifiers are biased for class A operation. A power amplifier may be class A, but higher power gain can be accomplished with the other classes: AB, B, and C. The main difference between a class A voltage amplifier and a class A power amplifier is the magnitude of the output voltage swing. Both of these circuits may amplify both voltage and power. Power gain is an incidental factor in voltage amplification, and the output has a relatively large swing. Voltage amplification may be accomplished in power amplifiers, but, if so, it is an incidental factor. The power amplifier has a relatively small signal swing.

### 8.2.1 Limits of Signal Swing

When the load line touches the power curve, the point of contact is always the center of the load line. This center point is also the operating point for maximum signal swing in a class A amplifier. Of course, the operating point can be set any place along the load line, but if it is not centered, a maximum swing results in either limiting or severe distortion. Consider the power amplifier circuit in Figure 8-4.

Power Amplifiers 181

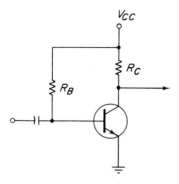

Figure 8-4. Class A Power Amplifier

This is a simple circuit with an *RC*-coupled input and a resistive collector load. The maximum power curve and load line are shown on the graph of Figure 8-5. The characteristics curves on this graph are exaggerated to illustrate the difference between theoretical and practical maximum signal swing. Theoretically, the signal can swing the full

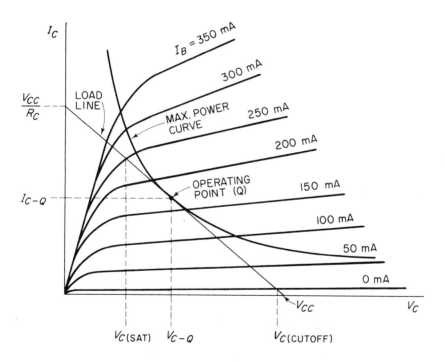

Figure 8-5. Load Line and Power Curve for Figure 8-4

length of the load line from $V_C = 0$ V to $V_C = V_{CC}$. In practice, the low limit of $V_C$ occurs when the load line crosses the curved portion of an $I_B$ line. This low limit of $V_C$ represents saturation. A swing beyond saturation results in intolerable distortion. The upper limit of $V_C$ occurs were the load line intersects the $I_B = 0$ line. This upper limit is cutoff, and no swing is possible beyond cutoff.

### 8.2.2 Power Across the Load

The maximum power across this transistor occurs at the operating point, point $Q$. At point $Q$, the transistor is idling with no signal input. The power at this point is

$$P = V_{C\text{-}Q} \times I_{C\text{-}Q}$$

On an actual load line, if point $Q$ is the maximum power point, $V_{C(\text{SAT})}$ and $V_{C(\text{cutoff})}$ occur at about equal distances from $Q$. This allows equal signal swing in both directions before distortion becomes serious.

When the signal is a perfect sine wave, point $Q$ will not shift. The average power across the transistor is

$$P_{AV} = \frac{V_{CC}^2}{4R_C}$$

The average signal power developed across the collector load resistor is the product of the RMS current and voltage. This average power is also

$$P_{R_C} = \frac{V_{CC}^2}{8R_C}$$

These two equations indicate that the maximum power that can be delivered to the load is equal to half the maximum power that the transistor can safely handle.

This type of audio power amplifier is very inefficient in terms of the power that must be furnished by the power supply as compared to the power delivered to the collector load resistor. The efficiency is about 25 percent.

## 8.3 Class A Amplifier with a Transformer Load

Efficiency of a class A power amplifier can be improved by using a transformer instead of the collector load resistor. The inductive load

# Power Amplifiers

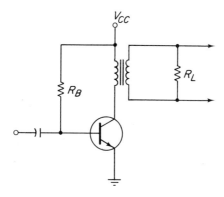

Figure 8-6. Transformer Load

reduces the dc resistance in the circuit and diminishes the dc power requirement on the power supply.

### 8.3.1 Sample Circuit

The use of an output transformer is illustrated in Figure 8-6. The resistor, $R_L$, can be the input to the next stage or simply a load on the transformer secondary to broaden the frequency response. Either way it reflects a resistance into the transformer primary. The reflected resistance is

$$R = \left(\frac{N_P}{N_S}\right)^2 R_L$$

This reflected resistance is an ac component, but there is also a dc load on the circuit.

### 8.3.2 Load Lines

This circuit needs an ac load line and a dc load line. The two load lines and the power curve are illustrated on the graph in Figure 8-7. The dc load line is constructed by assuming a zero resistance for the primary winding of the transformer. This produces a collector voltage equal to $V_{CC}$ for one end of the dc load line. Any value of $V_{CC}$ and zero resistance gives an $I_C$ of infinity for the other end of the line. The dc load line then becomes a vertical line from $V_{CC}$. The operating point, $Q$, is the intersection of the dc load line and the power curve.

The ac signal causes an inductive action and results in a signal swing from 0 V to $2 V_{CC}$. At the same time, $I_C$ swings from 0 to $2 I_{C\text{-}Q}$. The ac

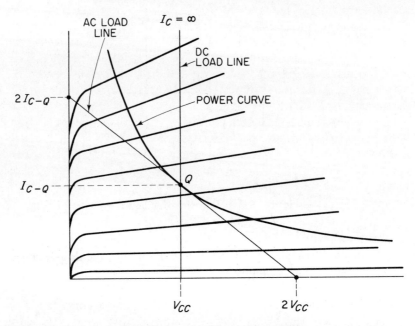

Figure 8-7. Ac and dc Load Lines

load line connects points $2\,V_{CC}$ to $2\,I_{C\text{-}Q}$. This line runs through the operating point and touches the maximum power curve.

The dc resistance in the primary of a transformer is *not* zero. In fact, this dc resistance is usually large enough to be reckoned with. With a considerable dc resistance in the primary, the dc load line is not vertical but will be tilted toward the upper left of the graph.

## 8.4 Distortion in a Class A Power Amplifier

A primary goal in designing an audio-frequency amplifier is a faithful reproduction of the input signal. This output signal should be directly proportional to the input signal in all respects without regard to the polarity, frequency, or amplitude of the input signal. This of course is an ideal output that is difficult to attain, but any deviation from this ideal is distortion. Class A audio-frequency power amplifiers are subject to two types of distortion: harmonic and intermodulation.

### 8.4.1 Harmonic Distortion

When the input signal to a class A power amplifier is composed of a single frequency, the output signal may contain the fundamental and several harmonics of the fundamental frequency. These harmonics alter the waveshape of the output signal; this is harmonic distortion. The nonlinear transfer characteristics of a transistor will generate harmonic components of the signal. Figure 8-8 illustrates this nonlinear function. The input is a perfect sine wave, one frequency; the output is distorted. The output contains the fundamental frequency and several harmonics of the fundamental. The amplitudes and phases of the harmonics have fixed values in respect to the fundamental. Only the amplitudes of these harmonics are considered here. The number of harmonics in the output is determined by the curvature of the transfer characteristics curve. The curve in Figure 8-8 is rather severe, and the output contains a large number of harmonics. The greater the curve, the larger the number of harmonics and the greater the distortion.

### 8.4.2 Intermodulation Distortion

An audio signal, except for a single pure tone, is not a pure sine wave. The standard audio signal is a complex wave containing many frequencies and constant frequency changes. This complex signal tends to cause intermodulation distortion in an audio power amplifier. The net result of intermodulation distortion is a mixing of two or more frequencies which produces sum and difference frequencies. This type of distortion is also related to the curvature of the transfer characteristics of the transistor.

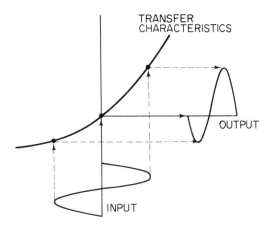

Figure 8-8. Harmonic Distortion

### 8.4.3 Correcting Distortion

It is difficult to predict the exact distortion that will be produced by a particular circuit. The characteristics curves are made for particular types of transistors rather than for each particular transistor. The current–voltage curves are also dependent on temperature and tend to vary with a change of power. These problems in analysis can be countered by experimentation.

Introducing degenerative feedback in the power amplifier stage tends to reduce the distortion. However, this feedback must be conservative. Too much degeneration not only eliminates gain but may also introduce limiting, which is a type of harmonic distortion.

A good procedure is to build the circuit with a small degenerative feedback. Design the circuit to deliver the required power and match the impedances. Then test for distortion. One to 2 percent distortion can be tolerated. More than that indicates that circuit modification is required. Adjust bias and feedback until the distortion reaches a minimum value. The adjustments will shift the operating point, alter the power output, and cause slight power loss because of mismatched impedances. However, all these changes should be very minor, and the overall performance should remain within a reasonable tolerance.

## 8.5 Class B Power Amplifier

Since the class B amplifier conducts only 50 percent of the time, it can deliver a much higher output power during its conduction time. However, a single class B amplifier would be of little use in audio power amplification. This small problem is easily overcome by connecting two matched transistors in a class B push–pull arrangement.

### 8.5.1 Sample Circuit

One push–pull circuit arrangement is illustrated in Figure 8-9. This is a traditional arrangement, and variations of it have been used for many years.

The active devices are a matched pair of *NPN* transistors arranged so that $Q1$ conducts on one alternation of the input, and $Q2$ conducts on the other alternation. This arrangement must be fed by identical inputs that are 180 degrees out of phase with each other.

The out-of-phase inputs can be taken from a phase-splitter driver stage or from a center-tapped transformer. In this case, a transformer is

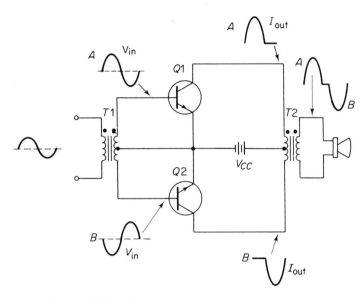

Figure 8-9. Class B Push–Pull Power Amplifier

used. Taking signals from both ends of the center-tapped secondary of $T1$ provides identical, but out-of-phase, inputs for the two transistors.

The positive alternation of signal A causes $Q1$ to conduct while the negative alternation of signal B holds $Q2$ cutoff. Each transistor amplifies one alternation of the input. The transformer, $T2$, puts the alternations back together to obtain the amplified sine wave to drive the speaker.

### 8.5.2 Distortion in Push–Pull Amplifiers

Notice that the output in Figure 8-9 *is not* a perfect sine wave. The distortion near the zero crossover point is caused by the fact that one transistor cuts off just as the other starts to conduct. This causes both transistors to operate on the nonlinear portion of its characteristics curve when its conduction is very low. This type of distortion is often much more severe than indicated in the drawing.

There are two ways to eliminate crossover distortion. Both methods are quite simple. The first method is to place a small forward bias across the emitter–base junction of both transistors. This bias changes each transistor to class AB operation. This AB is discussed in detail in Section 8.6.

The second way to eliminate crossover distortion is to use unipolar instead of bipolar transistors. The FET has linear characteristics that eliminate the distortion and still retain the class B operation for each transistor.

### 8.5.3 Complementary Class B Amplifier

The use of transformers, to a large extent, nullifies the benefits of solid-state design. Two of the prime objectives in solid-state design are light weight and compactness. Transformers are both heavy and bulky. Matched pairs of transistors can be arranged in a complementary fashion and used as push–pull power amplifiers. Such an arrangement is illustrated in Figure 8-10. This arrangement uses one *NPN* and one *PNP* with a common load resistor. Individually each transistor is a class B amplifier in a common-collector configuration. A single input signal operates both transistors. The positive alternation of the input places forward bias on $Q1$ and reverse bias on $Q2$. When $Q1$ conducts, electron current moves up through $R_L$, through $Q1$, to $+V_{CC}$. When $Q2$ conducts, electron current moves from $-V_{CC}$, through $Q2$, and down through $R_L$.

This class B arrangement also suffers from crossover distortion as evidenced by the output waveshape.

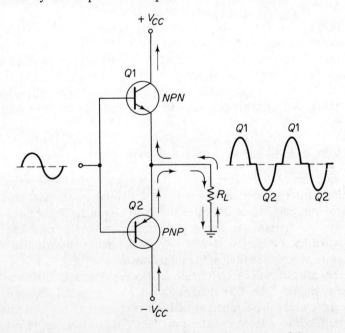

Figure 8-10. Complementary Class B Amplifier

## 8.6 Class AB Power Amplifier

As previously mentioned, the class B push–pull amplifier is easily converted to class AB. A class AB push–pull amplifier has a small forward bias on both transistors which causes each transistor to conduct slightly more than 50 percent of the time. This bias causes a slight current through both transistors during the crossover phase. The crossover distortion is effectively eliminated by keeping the transistors on the linear portion of their transfer characteristics curve.

### 8.6.1 Biasing with Transformers

One method of biasing a push–pull amplifier for class AB operation is illustrated in Figure 8-11. The biasing network for both transistors is composed of $R1$ and $R2$. The voltage drop across $R2$ is a very low value, normally about 0.75 V. This voltage across $R2$ is the bias voltage. Emitter resistors, $R3$ and $R4$, provide degenerative feedback. The value of $R3$ and $R4$ are equal, and neither value should exceed 1Ω. About 0.47 Ω is usually sufficient.

### 8.6.2 Biasing Complementary Circuits

There are two easy ways to forward bias complementary circuits. The first is voltage-divider bias as illustrated in Figure 8-12. Resistor $R2$ is

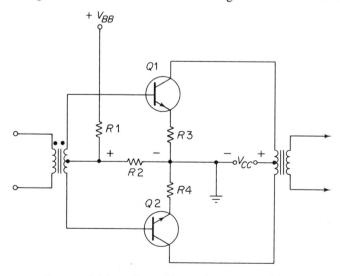

Figure 8-11. Class AB Push–Pull Amplifier

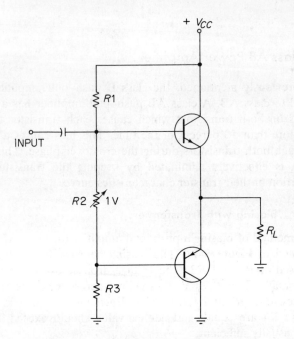

Figure 8-12. Voltage-Divider Bias

adjusted for a voltage drop of about 1 V across it. This provides enough bias to turn on both transistors slightly with no signal input. This arrangement places $R2$ in series with the input to $Q2$, but the disadvantage is not serious in most designs.

A more practical biasing method is to use a third transistor to provide bias for the complementary transistors, as illustrated in Figure 8-13. The current through $Q1$ provides the bias for both $Q2$ and $Q3$. The values of $R1$ and $R2$ determine the conduction level of $Q1$. These resistor values can be trimmed to eliminate most of the crossover distortion.

## 8.7 Design Example

A variety of designs are used in solid-state audio power amplifiers. Many of these designs utilize transformers despite their weight and bulk. The example used here is capable of operating a radio speaker. The specifications of the design problem are:

1. Signal input: 1 V (RMS) from a 500-Ω source.
2. Center frequency: 1000 Hz.

# Power Amplifiers

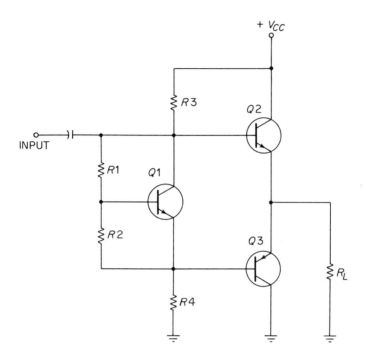

Figure 8-13. Transistor Bias

3. Speaker impedance: 8 Ω.
4. Required power: 5 W.
5. Available P.S. voltage: −22 V.
6. Transformer permissible.

## 8.7.1 Model Circuit

This is a low-power circuit. The design is simple, but care must be exercised to keep distortion to a minimum. The model circuit in Figure 8-14 should do the job. It is a single-ended amplifier using a bare minimum of components.

The input signal is $RC$-coupled to the base of $Q1$ by $C1$ and $R2$. The transistor has combination bias. Fixed bias is provided by the voltage divider consisting of $R1$ and $R2$. Self-bias is developed by $R3$. The effective input bias is $V_{R2} - V_{R3}$. Signal degeneration is prevented by bypassing $R3$ with $C2$. The collector load is the primary of an audio transformer, $T1$. The secondary of $R1$ delivers signal power to the speaker.

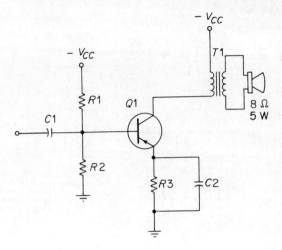

Figure 8-14. Model Circuit for a Power Amplifier

### 8.7.2 Selecting Components

The speaker requires 5 W of power. This 5 W of power must be delivered to an 8-$\Omega$ load. An audio transformer is selected with a secondary impedance of 8 $\Omega$. Transformers are rated according to primary impedance, secondary impedance, and output power capabilities. Transformers can be obtained, on special order, with exact primary to secondary relationships at specified power ratings. This special-order route is not practical in ordinary design. The transformer for this circuit should be an off-the-shelf item with a secondary impedance of 8 $\Omega$ and an output power rating of at least 5 W. A 6-W rating would be better. The extra power-handling ability will provide a margin of safety for extra reliability. Even a 10-W output capability would not be extravagant.

The primary of the transformer should match the output impedance of the transistor. The output impedance of $Q1$ can be determined by

$$Z_{\text{out}} = \frac{\Delta V_C}{I_C(\text{at } Q \text{ point})}$$

Since several factors are involved, some of these must be compromised. The best choice is to select a transformer with proper secondary impedance, proper output power rating, and the nearest available primary impedance. The current of $Q1$ can then be adjusted to approximately match the impedance of the primary. The relationship can be expressed as

$$I_C = \frac{2 \times P_C}{Z_{primary}}$$

The impedance of the secondary winding of $T1$ reflects into the primary according to the primary to secondary winding ratio. The reflected impedance is equal to the square of the turns ratio times the secondary impedance. In this case, the reflected impedance is

$$Z_P = \left(\frac{N_P}{N_S}\right)^2 \times Z_S$$

Considering the transformer to be 75 percent efficient, the primary power must be

$$P_P = \frac{P_S}{0.75}$$
$$= \frac{5 \text{ W}}{0.75}$$
$$= 6.67 \text{ W}$$

The RMS current is the square root of the power divided by the impedance. So RMS current in the primary is

$$I_{RMS} = \sqrt{\frac{P_P}{Z_P}}$$

The transistor must supply a peak current equal to 1.41 times the RMS value. Select a transistor capable of supplying twice as much current as this peak calculation reveals. Next, the values of $R1$, $R2$, and $R3$ may be estimated. These values must produce a bias that establishes the proper current through $Q1$. The procedure of obtaining the estimated resistance values is the same as described in previous amplifier circuits. The value of one or more of these resistors will probably need to be adjusted to obtain the proper quiescent current from $Q1$. When making this adjustment, be careful that distortion is not introduced into the stage.

The value of $C1$ should be such that it offers an $X_C$ equal to half the resistance of $R2$ at the lowest frequency to be amplified. At this audio frequency a 2-$\mu$F capacitor will do a good job.

The purpose of $C2$ is to bypass the ac signal around $R3$. This capacitor should be very nearly a short to audio frequencies. For most values of $R3$, $C2$ should be about 10 $\mu$F.

## 8.8 Troubleshooting

Trouble in the audio power amplifier appears in one of four ways: no output power, low output power, high output, or distorted output. Sometimes there will be a combination of two of these indications. The speaker should be included in troubleshooting of this stage, and the output power can be measured at the speaker.

### 8.8.1 No Output Power

No output power to the speaker is a symptom of an open circuit. The open can be in the input circuit, the transistor-biasing circuit, the transistor itself, in either primary or secondary of $T1$, or in the speaker. In all probability the open is a poor connection. The first step should be a careful inspection of all solder joints.

When all solder joints prove to be secure, an ac voltmeter should be used to take voltage measurements. It is assumed that a valid ac signal is coupled to the input of $C1$. Measure the voltage across $R2$. A zero reading means no input signal; $C1$ is open. A valid reading across $R2$ shows that $Q1$ has an input. Check the output from collector to ground. A zero output means that the open is in either the biasing circuit or in the transistor.

With no output signal from $Q1$, the trouble is very likely an open in either $R1$, $R2$, or $R3$. Remove the power from the circuit and check each resistor with an ohmmeter. Be sure to disconnect one end of the resistor before measuring the resistance. If all resistors are good, replace the transistor.

When there is a valid output from $Q1$, the trouble must be in either the transformer or the speaker. Remove the power and use the ohmmeter to check for continuity. No continuity in the primary winding means an open primary. Continuity in the primary winding means the open is in either the secondary winding or the speaker.

When the trouble has been isolated beyond the transformer primary, disconnect one lead between the secondary and the speaker. Check the secondary winding for continuity. No continuity in the secondary winding means an open secondary. Continuity across this winding indicates that the open is in the speaker.

## 8.8.2 Low or High Output Power

A power reading at the speaker which shows an output that is either too low or too high indicates improper bias on $Q1$. This can be corrected by changing the resistance of one or more of the biasing resistors ($R1$, $R2$, or $R3$).

An open $C2$ could cause a low output but not a high. If the output is low, disconnect one side of $C2$, and check it with an ohmmeter before altering the bias resistances. A shorted $C2$ could cause a high output but not a low. If the output is high, check $C2$ for a short.

## 8.8.3 Distorted Output

A distorted output can be caused from a bad speaker or from an improper operating point on $Q1$. Use an oscilloscope to check the output signal of $Q1$. Since a normal audio signal is a rather complex waveshape, use a 1000-Hz signal from a signal generator as the input signal. If the collector output is distorted, the transistor bias must be readjusted. It may be necessary to introduce degenerative feedback in order to minimize the distortion. This can be done by removing $C2$ or by adding another unbypassed resistor in series with $R3$. Either action will alter the operating point and require a readjustment of the bias. An undistorted signal on the collector isolates the trouble to the speaker.

CHAPTER NINE

# Noise in Audio Amplifiers

In electronic circuit design, any undesirable signal is classified as noise. Some of the common sources of electronic noise are electrical storms, ac power supplies, arcing in electrical devices, and induction from nearby circuits. In various stages of design it is important to plan filters to greatly reduce this external noise.

In audio circuits, the noise generated in the actual circuit components is of primary concern. When constructing a chain of several audio amplifiers, it is absolutely essential to minimize the noise in the first stage. Any noise generated in the first stage will be amplified through all subsequent stages.

## 9.1 Signal-to-Noise Ratio

The *signal-to-noise ratio* is a comparison of the signal strength of an intelligence signal to the strength of the noise that is carried with the signal. In audio circuits, any noise at all is objectionable; so only the highest possible signal-to-noise ratio is acceptable.

### 9.1.1 Ratio Expressions

Signal-to-noise ratios may be expressed as power ratios, voltage ratios, or decibels. When a ratio is expressed in decibels, an addition of each 3 dB doubles the power, and each 6 dB doubles the voltage. The decibel power ratio is defined as 10 times the ratio of two powers. The decibel voltage ratio is defined as 20 times the ratio of two voltages. The equations for both voltage and power signal-to-noise ratios are

$$dB = 10 \log \frac{P_{so}}{P_{no}} = 20 \log \frac{V_{so}}{V_{no}}$$

where $P$ = power, $V$ = voltage, $s$ = signal, $n$ = noise, and $o$ = output.

The signal-to-noise ratio in terms of either voltage or power may be expressed as a simple ratio such as 2:1, 4:1, 10:1, and so on. In this case, the first number represents signal strength and the second number represents the relative strength of the noise. If the ratio is in terms of power, a ratio of 4:1 indicates the power of the signal is four times the power of the noise. This is equivalent to a dB rating of 5 dB. The same rating (4:1) for voltage means a voltage four times as strong as the noise and is equated to a dB rating of 12 dB and a power rating of nearly 16:1. The table in Figure 9-1 compares several examples of the three ratio expressions.

### 9.1.2 Measuring Ratios

The easiest way to measure a signal-to-noise ratio involves the use of a signal generator and an oscilloscope. The signal generator is coupled to the input of an amplifier, and the oscilloscope is set to measure the voltage amplitude. This measurement will reveal the output signal amplitude. The connections are illustrated in Figure 9-2. This measurement provides one of the two readings for determining the voltage signal-to-noise ratio. Record the peak-to-peak voltage amplitude reading.

The next step is to determine the noise level under normal conditions with no signal input. The oscilloscope connections remain the same, but the signal generator is removed from the input. Instead of an

| POWER RATIO | VOLTAGE RATIO | DECIBEL RATIO |
|---|---|---|
| 1:1 | 1:1 | 0 |
| 2:1 | 1.5:1 | 3 |
| 4:1 | 2:1 | 6 |
| 8:1 | 3:1 | 9 |
| 16:1 | 4:1 | 12 |
| 100:1 | 10:1 | 20 |
| 10,000:1 | 100:1 | 40 |

Figure 9-1. Comparison of Signal-to-Noise Ratio Expressions

input signal, a resistance equal to the output resistance of the signal generator is connected in parallel to the input of the amplifier. This arrangement is illustrated in Figure 9-3.

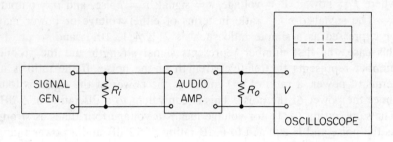

Figure 9-2. Measuring the Signal Voltage Amplitude

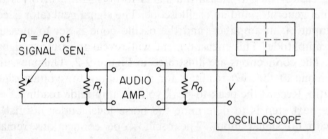

Figure 9-3. Measuring the Noise Level

The arrangement illustrated here may be referred to as a shorted input. But it is readily apparent that an actual short across the input would alter the bias and shift the operating point.

The peak-to-peak noise amplitude measurement is taken using the arrangement illustrated in Figure 9-3. This is the second reading for determining the signal-to-noise voltage ratio. The next step is calculation. The ratio is signal amplitude divided by noise amplitude:

$$\text{voltage ratio} = \frac{\text{signal amplitude}}{\text{noise amplitude}}$$

## 9.2 Noise Spectrum

The designer of audio circuits is primarily concerned with noise that falls in the audio-frequency band; however, audio signals can be hampered by noise at frequencies both above and below the audio range. In general terms, all noise can be placed in one of three categories: white noise, blue noise, and pink noise. The three categories are represented on the graph in Figure 9-4.

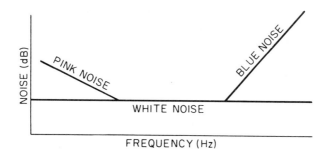

Figure 9-4. Noise Spectrum

### 9.2.1 White Noise

A *white noise* signal is a completely random signal. Its amplitude and frequency are completely unpredictable at any given point in time. Further, it delivers identical power at all frequencies in the entire spectrum. White noise is represented by the horizontal bar in Figure 9-4.

The white noise produced by a transistor can be of three types, and the type indicates how it is produced. These three types are thermal noise, shot noise, and partition noise.

*Thermal noise* is generated by the random motion of particles in a conductor caused from heat energy imparted to these particles. The bias resistors in the input circuit and the base resistance of bipolar transistors are important sources of thermal noise. These sources are of primary concern for the simple reason that any noise generated in the input circuit will be amplified through the transistor.

*Shot noise* is a type of white noise that is generated in a transistor junction. This noise is generated by the random passage of charged particles through a junction. Shot noise is directly proportional to the quantity of current through the transistor.

*Partition noise* is white noise generated in a bipolar transistor due to the division of emitter current between base and collector. This partition noise is inversely proportional to the beta of the transistor. The larger the beta, the smaller the base current and the smaller the noise. Obviously the easiest way to keep partition noise to a minimum is to utilize transistors with a very high beta.

### 9.2.2 Blue Noise

*Blue noise* occurs at the high end of the frequency spectrum, and for this reason it has little bearing on the design of audio circuits. Blue noise has an increase in power level that is directly proportional to an increase in frequency. In fact, the blue noise power increase is about 6 dB per octave. Blue noise is generated due to the rolloff of the transistor beta at high frequencies. The blue noise increases at the same rate that beta decreases.

### 9.2.3 Pink Noise

At low frequencies, a type of noise is encountered that is inversely proportional to frequency. This low-frequency noise is *pink noise*. Pink noise also has a constant energy factor; its power changes at the rate of 3 dB per octave. This low-frequency noise is caused from surface phenomena and leakage currents.

Pink noise is particularly bothersome in bipolar transistors. It is first noticed at about 1000 Hz and becomes worse as frequency decreases. It is first noticed in unipolar transistors at about 100 Hz.

A reduction of pink noise can be accomplished by minimizing the current through, and the voltage across, the transistor. Since leakage currents are affected by heat, the pink noise factor can be reduced by holding the transistor temperature to a minimum value.

## 9.3 Noise Factor

The *noise factor* should not be confused with the noise figure, even though both confer the same information. When expressed as a voltage or power ratio, it is a noise factor. When expressed in decibels, it is a noise figure. A power noise factor of 2, or 2:1, is equivalent to a noise figure of 3 dB. This figure, or factor, expresses the quality of an amplifier with respect to the noise; the lower the noise factor, the better the amplifier. A data sheet may express the noise figure as spot noise at some specified frequency. A more meaningful expression is a noise figure over the entire frequency band.

The noise factor is a ratio between two signal-to-noise ratios. The input signal-to-noise ratio of an amplifier is higher than the output signal-to-noise ratio because the transistor contributes to the noise. The noise factor is the input signal-to-noise ratio divided by the output signal-to-noise ratio:

$$F = \frac{S_{in}/N_{in}}{S_{out}/N_{out}}$$

Since the transistor's noise contribution is proportional to gain, the noise factor may be expressed as noise out divided by noise in times the gain of the stage:

$$F = \frac{N_{out}}{N_{in} \times \text{GAIN}}$$

The noise factor of an amplifier is affected by the operating point of the transistor, the output resistance of the signal source, and the frequency of the signal.

### 9.3.1 Operating Point

The noise factor of a particular transistor is directly proportional to both the emitter current and the collector voltage. This relationship is illustrated on the graph in Figure 9-5. The quiescent values of emitter

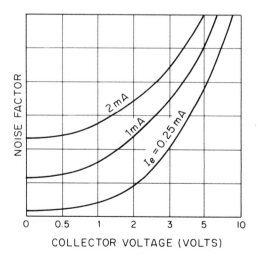

Figure 9-5. Noise Factor Versus Emitter Current and Collector Voltage

current and collector voltage are determined by the selection of the operating point. The operating point is set by selecting circuit values to produce a desired level of emitter current and collector voltage. Figure 9-5 indicates that the best noise factor for this transistor is obtained at a minimum of both collector voltage and emitter current. This is generally true of all transistors. Of course, transistors with zero current are serving no useful purpose, so designers must keep this graph in mind and set the operating point as low as possible and still do the job. In any case, it is a trade-off between noise level and current–voltage.

### 9.3.2 Source Resistance

The output resistance of the signal source plays a strong role in the noise factor of an amplifier. Figure 9-6 is a graph of noise factor plotted against the source resistance. This graph shows that the noise factor decreases as source resistance increases until the source resistance reaches about 800 Ω. At that point, the trend reverses and the noise factor increases with an increase in source resistance. A satisfactory noise factor for this transistor is obtained with a source resistance between 100 and 3000 Ω.

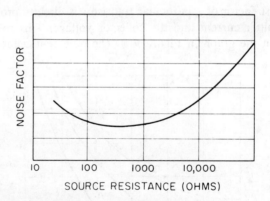

Figure 9-6. Noise Factor Versus Source Resistance

### 9.3.3 Signal Frequency

The frequency of the signal being amplified is also influential on the noise factor. This fact is illustrated in Figure 9-7. With this transistor the noise factor decreases with an increase of frequency up to about 50 kHz. Above 50 kHz, the noise factor increases rather sharply with an

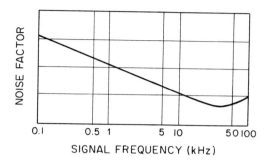

Figure 9-7.  Noise Factor Versus Signal Frequency

increase in signal frequency. This upturn in noise factor has no bearing in audio circuit design because it occurs at frequencies above the audio band. This relationship of frequency and noise factor is important. It enables the selection of a transistor with a noise factor that is compatible with the frequency that must be amplified.

## 9.4  Equivalent Noise Generators

The noise in an amplifier is produced by three noise generators. One of these noise generators is composed of the components of the input circuit. It generates noise and transfers it into the transistor. The remaining two noise generators are within the transistor. One of these internal generators produces voltage noise, and the other produces current noise. Figure 9-8 is an equivalent circuit of the two internal noise generators.

The voltage noise is amplified through the transistor. So voltage noise out is equivalent to voltage noise in multiplied by voltage gain. This voltage noise out can be measured by shorting the input. The cur-

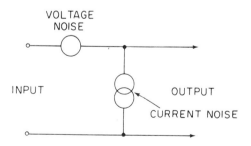

Figure 9-8.  Internal Noise Generators

rent noise in is determined by comparing the output current, with a known signal input, to the output noise when the input is open. The best possible noise factor is obtained when the resistance of the signal source is equal to voltage noise divided by current noise.

## 9.5 Bipolar Transistor Noise

All the noise information discussed thus far applies to bipolar transistors, but there are still other noise characteristics that are *peculiar* to bipolar devices. Also, a bit more elaboration on how to obtain the best noise factor seems to be essential. Perhaps the most important among these are source resistance and collector current.

### 9.5.1 Contour Curves

The variations of the noise factor, or the noise figure, with both collector current and source resistance can be described by a family of *contour curves*. Such a family of curves with the noise figure expressed in decibels is illustrated in Figure 9-9. These curves are useful in audio circuit design because they supplement the data normally given on data sheets. The normal noise data provided are the characteristics at a specified value of collector current, collector voltage, signal frequency, and source resistance. Of course, these specified conditions can never be duplicated in a practical circuit. The curves enable the designer to build a practical circuit to deliver the least possible noise for that particular job.

The objective is to construct the circuit to operate within the lowest possible noise factor contour and still do the job for which it was designed. This is accomplished by manipulating values of source resistance and collector current so that they intersect in the lowest possible contour. This will never be the ideal noise factor, but it will be the best possible under the circumstances.

### 9.5.2 Design Applications

The first step in designing a low-noise amplifier is the selection of the amplifying device. The source resistance is often dictated by circuit requirements. Examination of curves for several transistors enables the designer to select one that generates the least noise for reasonable quantities of collector current. Sometimes circuits are designed when only specified transistors are available. In this case, the curves enable circuit designers to place the operation within a reasonably low-noise-figure contour.

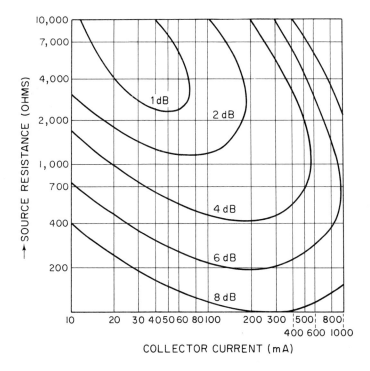

Figure 9-9. Noise-Figure Contour Curves

When using bipolar transistors in the design of audio-frequency amplifiers, it is a good practice to keep collector voltage, collector current, and source resistance to the minimum practical values. If this rule is followed carefully, it will assure the best noise factor under the operating conditions.

## 9.6  Field-Effect-Transistor Noise

Some signal sources have high output resistance. When designing circuits to work with these devices, designers must contend with what they have. One example of a high-resistance signal source is a crystal pickup head. With this and other high-resistance sources, the designer must contend with the high resistance. If he uses a bipolar transistor, he must also contend with a high noise factor. Fortunately, the field-effect transistor is a high-impedance input device, and its noise factor in respect to source resistance is drastically different from that of the bipolar transistor. The source resistance has an influence on noise but a different result. Other factors that affect FET noise are signal frequency and drain-to-source voltage.

### 9.6.1 Source Resistance

Since the FET has a very high input impedance, a high source impedance not only provides easy impedance matching, but it also lowers the noise factor. This noise factor–resistance relationship is illustrated in Figure 9-10. This graph shows that the noise factor decreases with an increase in source resistance. This decrease is linear up to a resistance of about 600 k$\Omega$. At this point it begins to level off and bottoms out at a source resistance of 1 megohm (M$\Omega$). As source resistance increases beyond 1 M$\Omega$, the noise factor increases very slowly and at a linear rate.

So with an average FET, the best noise factor can be obtained with a source resistance of about 1 M$\Omega$. The noise factor is still very low, with a source resistance of 10 M$\Omega$.

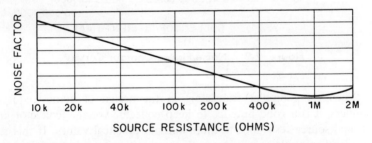

Figure 9-10. FET Noise Factor and Source Resistance

### 9.6.2 Signal Frequency

This is another area where the FET is superior to the bipolar transistor for audio-frequency amplification. Examine the graph in Figure 9-11. This graph is for the low end of the audio-frequency band. As the signal frequency dropped from the top of the audio band to 4 kHz, the noise factor held constant and at the very low level illustrated here. As signal frequency decreases further, the noise factor holds that low level down to about 300 Hz. Now it starts a slow increase which reaches a linear rate at about 125 Hz. This is the pink noise area for the FET. The higher the transconductance, the lower the pink noise.

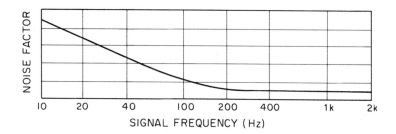

Figure 9-11.  FET Noise Factor and Signal Frequency

### 9.6.3  Drain-to-Source Voltage

There is a direct linear relationship between the noise factor and the drain-to-source voltage. This relationship is illustrated in Figure 9-12. Even though this is clearly a linear ratio, the noise factor remains very low for all reasonable values of drain-to-source voltage. The maximum power level of the noise on this graph is slightly less than 1 dB.

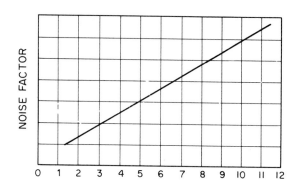

Figure 9-12.  FET Noise Factor and Drain-to-Source Voltage

**CHAPTER TEN**

# Multistage Amplifiers

The principal difference between the design of single-stage amplifiers and *multistage amplifiers* is the design of the interstage coupling. Three types of interstage coupling are commonly used in solid-state audio amplifier circuits: resistor–capacitor (RC), transformer, and direct. None of these are perfect and none are best suited to all situations. Each type has advantages and disadvantages that need to be weighed in order to best serve a given situation.

The gain of an individual amplifier stage is limited by four factors: transistor parameters, impedance of the input circuit, the load that it must feed, and the biasing circuits. The biasing circuits include bias-stabilization circuits. In multistage amplifiers, the signal output of one stage becomes the input signal to the next. The overall circuit gain then becomes the product of the gains of the individual stages.

## 10.1 Resistor–Capacitor Coupling

*Resistor–capacitor coupling* is probably the most frequently used of the three types. A capacitor is a convenient path for a signal between stages, and it serves a useful purpose by confining direct voltages to individual stages. Further, the *RC* coupling can be arranged to form either high-pass or low-pass filters. These advantages make *RC* filters both adequate and desirable for many coupling applications.

On the other hand, capacitors block dc and attenuate very low frequencies. This action can be a disadvantage in some cases. In high-fidelity audio work, there is another objectional feature to *RC* coupling; it has a frequency-dependent phase shift across the capacitor.

### 10.1.1 RC Coupling as a High-Pass Filter

*RC* coupling can be formed into two types of filters: *high-pass filter* and *low-pass filter*. The high-pass filter discriminates against low frequencies and freely passes signals above a specified frequency. Figure 10-1a illustrates the high-pass filter, while Figure 10-1b shows the approximate response curve for this type of filter. Frequency $f_o$ is the corner frequency where attenuation of the signal begins. Signals at frequencies above $f_o$ are passed without attenuation. Signals at frequencies below $f_o$ are attenuated at the rate of 6 dB per octave or 20 dB per decade. The corner frequency $f_o$ is determined by

$$f_o = \frac{1}{2\pi RC}$$

where the resistance is in ohms and the capacitance is in farads.

Two or more high-pass filter sections may be connected in series to increase the rate of rolloff beyond the 6 dB-per-octave rate. The reduction in gain is equal to the sum of the reduction of all the filter sections.

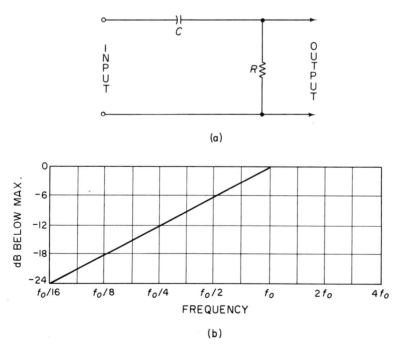

Figure 10-1. High-Pass RC Filter and a Frequency-Response Curve

### 10.1.2 RC Coupling as a Low-Pass Filter

The low-pass *RC* filter passes signals at a low frequency and discriminates against signals that are above a specified frequency. Figure 10-2a illustrates a low-pass filter, while Figure 10-2b shows its approximate frequency-response curve. Again the corner frequency is $f_o$. Signals at frequencies below $f_o$ are readily passed. Signals at frequencies above $f_o$ are attenuated at the rate of 6 dB per octave. The same equation used for high-pass corner frequency can be used for the low-pass corner frequencies, that is,

$$f_o = \frac{1}{2\pi RC}$$

where resistance is in ohms and capacitance is in farads.

Two or more low-pass filter sections can be connected in series to increase the rate of rolloff for frequencies above $f_o$. The reduction in gain is equal to the sum of the reductions of all the filter sections.

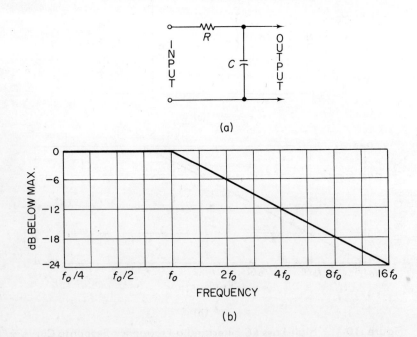

Figure 10-2. Low-Pass RC Filter and a Frequency-Response Curve

### 10.1.3 Bandpass Filter Circuits

The *RC* circuit effectively becomes a *bandpass filter* when both a high-pass and a low-pass filter are connected in the same circuit. The same result can be obtained with an amplifying stage between the two filters. In fact, the filtering results are more predictable when the two filters are separated by an amplifier. Figure 10-3 is an approximate frequency-response curve for a bandpass filter. The corner frequency for the low-pass filter is $f_o1$, while $f_o2$ is the corner frequency for the high-pass filter. Both of these $f_o$ points are located by the equation previously used. The effective output here is indicated as being 12 dB down from maximum, which is fairly realistic. The effective bandpass extends between a point 3 dB down from $f_o2$ and another point 3 dB down from $f_o1$.

### 10.1.4 Equivalent Filter Circuits

The input circuit to a transistor is in reality an *RC* filter, whether or not it is the designer's intention. The *equivalent circuit* for such a filter is shown in Figure 10-4a, while Figure 10-4b is its approximate frequency-response curve. Resistor $R_s$ is the equivalent output resistance of the previous circuit. Resistor $R_p$ is the parallel resistance at the input to the transistor, which may be either a physical resistor or the input resistance of the device. Capacitor $C$ is the input capacitance of the transistor.

The maximum output from this circuit is somewhat less than zero dB because of the resistance in the circuit. Even though this is essentially a low-pass filter, the low frequencies, below $f_o$, have some attenuation

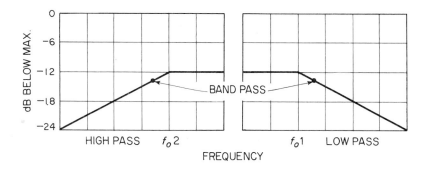

Figure 10-3. Bandpass Filter Frequency-Response Curve

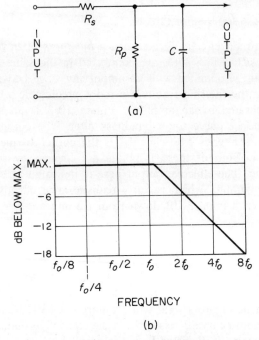

Figure 10-4. Transistor Input Filter and a Frequency-Response Curve

due to the voltage division between $R_s$ and $R_p$. The actual voltage reduction depends upon the relative values of $R_s$ and $R_p$:

$$V_{out} = \left(\frac{R_p}{R_s + R_p}\right) V_{in}$$

At frequencies below $f_o$, $C$ is an open circuit and has no effect on the output voltage. The attenuation in terms of dB can be obtained directly by

$$dB = 20 \log_{10}\left(\frac{R_p}{R_s + R_p}\right)$$

The corner frequency $f_o$ is a bit more involved. The equation can be stated as

$$f_o = \frac{1}{2\pi R_e C}$$

where $R_e$ is the equivalent resistance of $R_s$ in parallel with $R_p$.

## 10.1.5 Practical Application

In Figure 10-5, a practical, often-used $RC$ circuit couples the signal from the output of $Q1$ to the input of $Q2$. The physical $RC$ coupling here is $C1R2$. However, since the resistance of $R2$ is probably more than 10 times that of the input resistance of $Q2$, $R2$ drops from the calculation. This leaves $R1$, $C1$, and $R_{in}$. ($R_{in}$ is the input resistance of $Q2$.) In effect, the input to $Q2$ is passed through $C1$ and taken across the combined resistance of $R1 + R_{in}$. This is a low-pass filter with a corner frequency of

$$f_o = \frac{1}{2\pi C1(R1 + R_{in})}$$

The $RC$ circuit in the emitter of $Q2$ also affects the frequency response of the stage. This is a high-pass filter. The signals at frequencies below $f_o$ are greatly attenuated. Starting at $f_o$, the gain increases at the rate of 6 dB per octave with no top limit. The corner frequency for this circuit is

$$f_o = \frac{1}{2\pi R4C2}$$

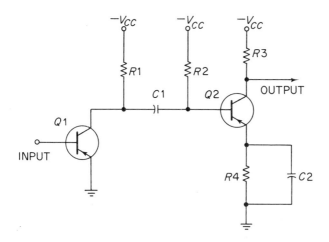

Figure 10-5. Practical RC Coupling

## 10.2 Transformer Coupling

Despite the miniature size of some modern coupling transformers, wide use of *transformer coupling* tends to counteract one main advantage of solid-state design: compactness. The smallest of transformers require as much space as a full-sized transistor, and it adds a small quantity of weight. In integrated circuits there is no place for transformers because printed circuit transformers are not practical devices. Despite these disadvantages, transformer coupling is still popular in standard solid-state circuits because of the ease in matching circuits of widely differing impedances. Other disadvantages of transformer coupling are discrimination against high frequencies and a frequency-dependent phase shift.

### 10.2.1 Efficiency

The percentage of efficiency for a transformer is computed by output power divided by input power. The efficiency may also be calculated using the resistance and the square of the turns ratio:

$$\% \text{ efficiency} = \frac{100}{1 + \left(\frac{R_p + N^2 R_s}{N^2 R_p}\right)}$$

where $N$ is the turns ratio, $N_p/N_s$, $R_p$ is the resistance of the primary, and $R_s$ is the resistance of the secondary.

Some modern audio-frequency transformers are being constructed with an efficiency of more than 99 percent, but these are cost prohibitive except for very sophisticated design. The average off-the-shelf coupling transformer has an efficiency of about 75 percent. It is a good practice to keep this in mind and to allow a 25 percent loss factor to compensate for this inefficiency.

### 10.2.2 Reflected Impedance

One often-confusing aspect of a transformer is its reflected impedance. Any load placed across the secondary winding will be reflected into the primary winding in accordance with the square of the turns ratio. A resistive load, which is most common in audio amplifiers, reflects a resistance. An inductive load reflects a capacitive reactance, and a capacitive load reflects an inductive reactance. The equation for reflected resistance is

$$R_r = N^2 R_s$$

where $R_r$ is reflected resistance, $N$ is the turns ratio $N_p/N_s$, and $R_s$ is the total resistance in the secondary circuit. The illustration in Figure 10-6 will further clarify this point. The number of turns in the primary winding is represented by $N_p$ while $N_s$ is the number of turns in the secondary winding. The turns ratio is $N$, and $N$ is found as follows:

$$N = \frac{N_p}{N_s}$$

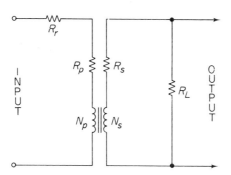

Figure 10-6.  Reflected Resistance

The small dc resistance of the primary winding is $R_p$ where $R_s$ is the same for the secondary winding. The load resistance placed across the secondary winding is $R_L$. The total resistance in the secondary circuit is $R_s + R_L$. The portion of this total resistance that is reflected to the primary is $R_r$.

The turns ratio of a transformer may not always be clearly indicated. This is a small problem. The turns ratio is always the same as the voltage ratio, and the voltage ratio is easily measured. Simply place an alternating voltage on the primary and measure both $V_p$ and $V_s$. The voltage ratio and the turns ratio can be obtained by dividing $V_p$ by $V_s$.

It is possible to have transformers constructed with precise values of impedance in both primary and secondary. Such a transformer makes it possible to make exact impedance matches. This is obviously an expensive way to build a circuit. A more practical way is to buy off-the-shelf transformers and compromise for less perfect impedance matches. It is possible to purchase transformers with exact values of secondary impedance. The primary should also be as close as possible, but a 25 percent mismatch is not too bad. Of course, the transformer ratings must be sufficient to withstand the currents and voltages of the circuit.

### 10.2.3 Transformer Coupling as a High-Pass Filter

The primary of the AF transformer serves as the output load impedance for one stage, and the secondary provides the input signal for the next stage. The equivalent of the primary circuit is illustrated in Figure 10-7a, and the approximate frequency-response curve is shown in Figure 10-7b. The frequency-response curve shows that the primary circuit of the transformer is a high-pass $RL$ filter. The $L$ is the inductance of the primary winding, while $R$ is the total impedance in the primary circuit. This $R$ includes physical resistance plus the reflected resistance from the secondary. This curve is identical to that of the $RC$ high-pass filter in Figure 10-1. The actions are essentially the same, but the corner frequency is probably different. The corner frequency here is determined by

$$f_o = \frac{R}{2\pi L}$$

At frequencies below $f_o$, the rolloff is at the rate of 6 dB per octave.

### 10.2.4 Transformer Coupling as a Low-Pass Filter

The secondary winding of the transformer is essentially a signal source with output taken across a resistor. The equivalent circuit is illustrated in Figure 10-8a, while 10-8b shows the approximate frequency-response curve. This of course is a low-pass $RL$ filter with actions similar to the $RC$ low-pass filter in Figure 10-2. The inductance $L$ is the inductance of the secondary winding, while $R$ is the load resistance placed across the secondary. This resistive load is always present, whether physical or otherwise. In many cases, the load across the secondary is the input of the transistor being fed by the transformer. The corner frequency for this low-pass filter is still

$$f_o = \frac{R}{2\pi L}$$

At frequencies above $f_o$, the rolloff is at the rate of 6 dB per octave.

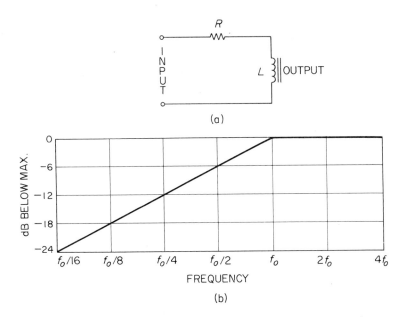

Figure 10-7. Primary Circuit and a Frequency-Response Curve

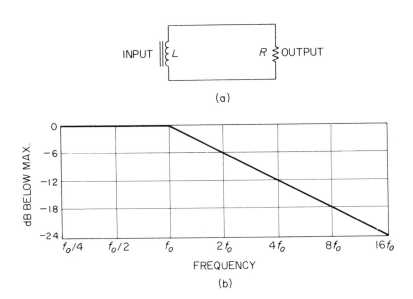

Figure 10-8. Secondary Circuit and a Frequency-Response Curve

### 10.2.5 Transformer Coupling as a Bandpass Filter

Since both windings of a transformer are required for coupling, the use of a transformer is, in effect, connecting a high-pass filter in series with a low-pass filter. Therefore, a transformer is a bandpass filter. The total approximate frequency-response curve is illustrated in Figure 10-9.

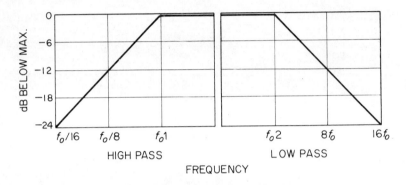

Figure 10-9. Transformer Bandpass Response Curve

This curve was created by combining the curves in Figures 10-7 and 10-8. The ideal operating area extends from 3 dB below $f_o1$ to 3 dB above $f_o2$.

### 10.2.6 Measuring the Transformer Frequency Response

Of course, the actual frequency-response curves are never as smooth as they appear on the approximate response curves. Despite the great care used in the construction of transformers, it is impossible to eliminate all the capacitance. The distributive capacitance between the windings, although very small, is always present. At some frequency, the capacitance and inductance are a resonant tank circuit. This resonant frequency and its harmonics cause peaks in the response curve. In most cases, these peaks occur at frequencies outside the critical AF range, but this is not always true.

The best way to be sure if a given transformer will perform a particular job is to measure its response over the entire range of frequencies that it is required to handle. Figure 10-10 illustrates a test circuit for measuring this frequency response. The total primary resistance, including that reflected from the secondary, is included in $R_{in}$. This means that, for reliable results, the value of resistances in both primary and secondary must be as near as possible to that of the actual circuit. In fact, the transformer frequency response can be measured with the transformer connected in the practical circuit.

The output of an AF signal generator is connected across a potentiometer. The potentiometer will need to be adjusted from time to time to keep a constant amplitude of input voltage. The signal generator is then tuned through the range of frequencies that the circuit is designed for. A complete and accurate frequency response curve can be constructed by plotting output voltage amplitude against frequency.

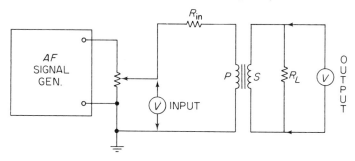

Figure 10-10. Measuring the Overall Frequency Response

### 10.2.7 Practical Application

The circuit in Figure 10-11 illustrates a practical application of transformer coupling between two stages of audio-frequency amplification. The impedance across the secondary winding of $T1$ is $R4$, plus the ac emitter–base resistance, multiplied by the beta of $Q2$:

$$R_L = \beta(R4 + r_e)$$

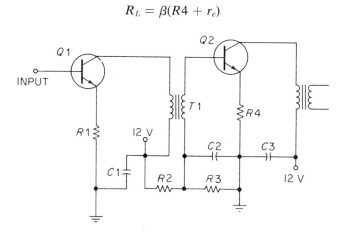

Figure 10-11. Practical Transformer Coupling

This secondary load resistance is reflected into the primary according to the square of the turns ratio. The reflected resistance is added to the resistance of $R1$ to form the resistance in the primary circuit. Capacitors $C1$, $C2$, and $C3$ are ac decoupling capacitors. Resistors $R2$ and $R3$ form a voltage-divider biasing circuit. The bias on $Q2$ is the difference between the voltage drops across $R3$ and $R4$. Degenerative feedback is provided by the unbypassed resistor $R4$.

Actual power requirements, and rated power capabilities, of a transformer must be carefully observed. The rated power capabilities should always be at least twice the actual power requirements of the circuit. This provides a margin of safety that can guard against expensive burnouts.

## 10.3 Direct Coupling

Many problems inherent in both *RC* and transformer coupling can be avoided by using *direct coupling*. Direct coupling is very popular in solid-state circuit design, and it is the most versatile of the three types. However, this type of coupling also has its own peculiar advantages and disadvantages, which prevent it from completely replacing the other types.

Direct coupling has no isolation components. This means that the direct voltage level at the output of one stage is the same as the level at the input of the next stage. This limits the number of successive stages that can be connected with direct coupling. Considerable quantities of feedback can be handled by direct coupling without causing low-frequency instability. Quiescent bias becomes a considerable problem in direct coupling because instability in any stage is amplified through all subsequent stages.

## 10.4 Using Discrete Components

The modern solid-state circuit designer has a choice of just how much of his circuit he will build with discrete components. Hybrid multistage amplifiers are available in packages at reasonable cost. When using discrete components the designer selects and tests each individual component.

### 10.4.1 Practical Circuit

Figure 10-12 is a sample of two discrete component amplifiers interconnected by direct coupling. This drawing illustrates two bipolar

# Multistage Amplifiers

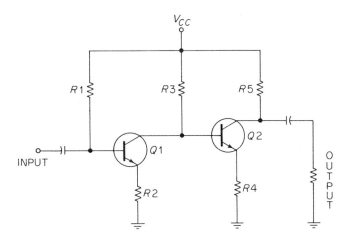

Figure 10-12. Using Discrete Components

transistor amplifiers. The collector output of the first stage is connected by a straight wire to the base of the second stage. Resistors $R1$ and $R2$ provide bias for $Q1$. Degenerative feedback for $Q1$ is provided by $R2$. The bias for $Q2$ is determined by $R3$ and $R4$. Resistors $R3$ and $R4$ also serve dual functions. The collector load for $Q1$, and base bias for $Q2$, are furnished by $R4$. Degenerative feedback, and emitter bias, for $Q2$ are provided by $R4$. The collector load for $Q2$ is resistor $R5$. The output is developed across $R5$, then coupled out through a capacitor. The effective base current for $Q1$ is $V_{cc}$ divided by the $(R1 + \beta R2)$:

$$I_B = \frac{V_{cc}}{R1 + \beta R2}$$

### 10.4.2 Second-Stage Bias

Providing both bias and stability for the second and subsequent directly coupled stages can be a tricky problem. A Thévenin equivalent circuit is helpful in planning the bias. This equivalent circuit for $Q2$ is illustrated in Figure 10-13. Battery $V_C$ represents the static collector voltage of $Q1$ when the two stages are completely isolated from each other. The total resistance looking into the base of $Q2$ is $R3 + \beta R4$. Beta, in this case, is the dc beta of $Q2$. The base current for $Q2$ is

$$I_B = \frac{V_C}{R3 + \beta R4}$$

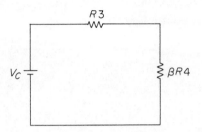

Figure 10-13. Equivalent Bias Circuit

Since $R4$ affects both bias and feedback, it exerts a drastic effect on the gain of the stage. If the resistor value is large enough to provide the desired bias, it may generate too much feedback. When such a problem exists, it can be resolved by connecting another resistor in series with $R4$. This added resistor value is such that the total emitter resistance provides proper bias. This added resistor is then shunted by a capacitor to prevent an increase in degenerative feedback. The bypass capacitor must be a short to the lowest frequency being amplified.

### 10.4.3 Stability

Bias drift is a serious instability problem in direct coupling. The stability factor for the circuit is the product of the stability of all the stages. A slight bias drift in the first stage is amplified through all stages. In Figure 10-12, the stability factor for $Q2$ is

$$\text{SF} = \frac{\beta R3}{\beta R4 + R3}$$

The stability factor for the two-stage circuit is the stability factor of $Q1$ multiplied by the stability factor of $Q2$.

## 10.5 Hybrid Packages

A variety of multistage, direct-coupled amplifiers are available in low-cost hybrid packages. Some of these are the phase inverter, differential amplifier, series output amplifier, complementary amplifier, emitter-coupled amplifier, and Darlington compound. These and many others are available in complete circuits and in various circuit combinations. Of course, the designer may still prefer to construct his circuits from discrete components, but it is difficult to match the packaged circuits in either cost or performance.

### 10.5.1 Darlington Compound

The *Darlington compound* is a useful audio-frequency power amplifier. It is frequently used as a driver to drive a final stage of power amplification. The circuit configuration is illustrated in Figure 10-14.

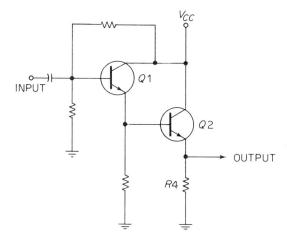

Figure 10-14. Darlington Compound

The circuit is essentially two common-collector amplifiers with the output of the first driving the second. The input impedance of this circuit is approximately equal to the product of beta of $Q1$, times beta of $Q2$, times $R_L$. Since the betas are usually matched, the input impedance can be expressed as

$$Z_{\text{in}} = \beta^2 R_L$$

The output impedance is approximately equal to the resistance of $R_L$. This high input and low output impedance is characteristic of common-collector circuits, and it enables matching a high impedance to a low impedance.

Since both transistors are in the common-collector configuration, the circuit has a voltage gain less than unity. The gain in both current and power is considerable.

### 10.5.2 Differential Amplifier

The *differential amplifier* is designed to produce an output only when there is a difference between the two inputs. They are particularly useful in eliminating common-mode signals. A common-mode signal, such as power-line pickup and other stray noise, drives both inputs equally, and the feedback removes, or greatly reduces, the signal. Figure 10-15 illustrates a circuit for a differential amplifier.

This circuit has a single output. This output is an amplified and inverted version of input 1 provided that it is different from input 2. If two outputs are desired, a collector load can be used and an output taken from $Q2$. This arrangement allows an inverted output of both inputs when they are different.

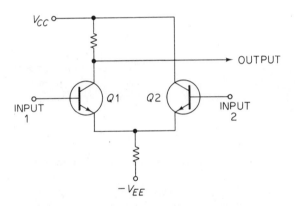

Figure 10-15.   Differential Amplifier

### 10.5.3   Series Output Amplifier

The *series output amplifier* is a push–pull amplifier frequently utilized to drive a speaker. The basic circuit is illustrated in Figure 10-16. The two inputs come from a phase splitter. Each input is developed alternately across $R_L$. Resistor $R_L$ can be replaced by a speaker coil. This dual-power-supply arrangement enables a direct connection to the load resistor without using a coupling capacitor. Eliminating the capacitor provides excellent low-frequency response. If very low frequencies are not an important consideration, the same result can be accomplished with one power supply. In this case, $-V_{EE}$ is replaced by a ground connection, and a coupling capacitor is placed in series with $R_L$.

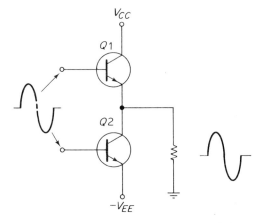

Figure 10-16. Series Output Amplifier

### 10.5.4 Complementary Amplifier

*Complementary amplifiers* can be obtained in quasi-complementary or full-complementary. When a complementary amplifier is used to drive a speaker, the necessity of a phase splitter is eliminated. The complementary amplifier also comes as either a voltage amplifier or a power amplifier. The full-complementary power amplifier circuit is illustrated in Figure 10-17. This circuit is composed of two Darlington compounds

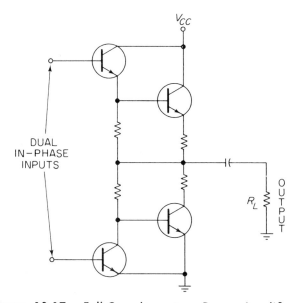

Figure 10-17. Full-Complementary Power Amplifier

and is commonly known as the dual Darlington output amplifier. It takes dual in-phase inputs and provides power amplification sufficient to drive a speaker.

The full-complementary voltage amplifier circuit is illustrated in Figure 10-18. This circuit is composed of two direct-coupled compounds. It is commonly called a direct-coupled output amplifier. This circuit takes dual in-phase inputs and provides voltage amplification.

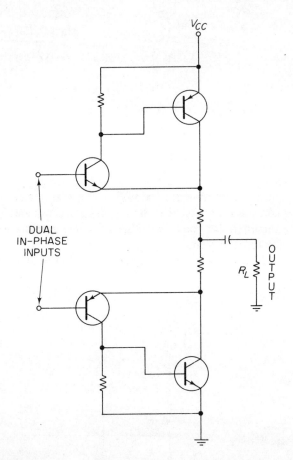

Figure 10-18. Full-Complementary Voltage Amplifier

The coupling capacitor in either of the complementary amplifier circuits can be omitted by using two power supplies. This can be done by removing the ground connection and connecting that point to a negative voltage source.

CHAPTER ELEVEN

# Audio Feedback

In most amplifier circuits the gain is largely dependent upon the characteristics of the amplifying device and the value of the circuit components. Proper attention to the design of feedback circuits can relieve some of this parameter dependency and enable a broader selection of transistors for a given application.

A *feedback circuit* senses the output and sends a specified portion of the output back to the input of this stage, or to some earlier stage, in the amplifier chain. The phase of the feedback is arranged so that it is either in phase or out of phase with the input. An *in-phase feedback* increases gain; an *out-of-phase feedback* decreases gain. The in-phase feedback is called *regenerative* or *positive feedback*. The out-of-phase feedback is *degenerative* or *negative feedback*.

In audio amplifiers, the feedback is utilized for the purpose of increasing the fidelity of the circuit. So in most audio circuits, the feedback is degenerative. The degenerative feedback improves fidelity at the expense of gain. The forward circuit amplification is the gain without feedback. With feedback, the gain remains relatively constant regardless of the forward circuit amplification.

Feedback is classified in several ways: voltage, current, compound, series, and parallel. The different types of feedback and the different methods of application lead to a variety of circuits with various characteristics.

## 11.1 Feedback Principles

Some thought must be devoted to the type and quantity of feedback that is required for best results in any given stage. In addition to the quantity of feedback, the designer must decide if it is to be degenerative or regenerative. In addition to the effect on the gain, he must be able to estimate the effect on the stability of the amplifier.

### 11.1.1 Feedback Equation

The input signal to the transistor proper is considered the actual input signal. This, in some cases, is less than signal amplitude at the output of the signal source. The difference between signal-voltage and voltage-in is illustrated in Figure 11-1. In this drawing, there are two resistors between the output of the source and the input to the amplifier. Therefore, the signal voltage at the input is less than the actual source voltage. The possible difference between these two amplitudes must always be considered.

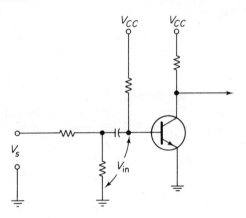

Figure 11-1. Signal Voltage and Input Voltage

The voltage amplification is the output voltage divided by the input voltage:

$$A_v = \frac{V_{out}}{V_{in}}$$

The voltage feedback circuit takes a fraction of the output voltage and couples it back to the input. This action is illustrated in Figure 11-2. The input voltage is now modified by the feedback voltage. Voltage gain with feedback becomes

$$A_{vf} = \frac{V_{out}}{V_{in} - FB}$$

where $V_{out}$ and $V_{in}$ are voltages without feedback. Since the ratio of $V_{out}$ to $V_{in}$ is voltage gain, the equation can be restated as follows:

$$A_{vf} = \frac{A_v}{1 - \text{FB}}$$

When the feedback is degenerative, the feedback is a negative quantity which indicates a 180-degree phase shift. Subtracting any negative quantity from one causes the feedback factor $1 - \text{FB}$ to become a quantity greater than unity. So gain with negative feedback is always less than gain with no feedback.

When the feedback is regenerative, the feedback is a positive quantity indicating no phase shift. Subtracting any positive quantity from one causes the feedback factor $1 - \text{FB}$ to become a quantity less than unity. So gain with positive feedback is always greater than gain with no feedback. When the feedback (FB) becomes $+1$, the denominator of the equation becomes zero. A zero denominator indicates a maximum amplification, and this constitutes an oscillating condition in the amplifier.

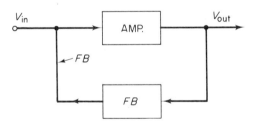

Figure 11-2. Amplifier with Feedback

### 11.1.2 Feedback for Stability

Instability to the point of causing oscillations is an intolerable condition in an audio-frequency amplifier unless that amplifier is specifically designed to be an oscillator. This statement holds true for all frequencies: not just those in the AF band. An audio driver with high-frequency oscillations can burn out speakers. An AF amplifier with high-frequency oscillations can cause sound distortion over the entire AF band. Any amplifier will oscillate at some frequency if it is driven hard enough. For good audio reproduction, designers must determine any tendency toward oscillation and eliminate this possibility from their circuits.

There is a simple experimental method to determine oscillation tendencies. It involves a test circuit and measurement of feedback over a wide range of frequencies. Suppose that the amplifiers under test are in the circuit of Figure 11-3. Resistors $R1$ and $R2$ form a voltage divider for developing the feedback. The total output voltage appears across the combination of $R1 + R2$. The portion of the output that appears across $R1$ is the feedback voltage. If $R2$ is disconnected from the circuit, the signal voltage across $R1$ still provides feedback for $Q1$. Since the signal potential across $R1$ opposes the action of the input signal, this is degenerative feedback. With $R2$ back in the circuit, part of the output from $Q2$ is now developed across $R1$. This feedback from $Q2$ also opposes the action of the input signal to $Q1$.

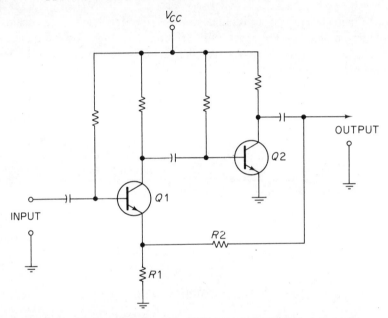

Figure 11-3. Amplifier with Feedback

The first type of feedback ($R2$ disconnected) is current feedback because it is produced from current through $Q1$. The quantity of feedback here is equal to $I_E \times R1$.

The second type of feedback is voltage feedback because $R1$ and $R2$ form a voltage divider. The level of this feedback voltage is

$$\text{FB} = V_{\text{out}} \left( \frac{R1}{R4} \right)$$

The level of this voltage feedback can be measured as illustrated in the test circuit of Figure 11-4. There is only one minor modification on the circuit. Resistor $R2$ has been disconnected from $R1$ and connected to $R3$. The value of $R3$ must be the same as that of $R1$. A signal generator provides the input signal, and the ac voltmeter across $R3$ is used to measure the feedback.

The signal generator is now tuned through a range of frequencies that includes at least all frequencies that the amplifier is expected to handle. Care must be exercised to keep the amplitude of the input at a constant value for all frequencies. The output is measured across $R3$.

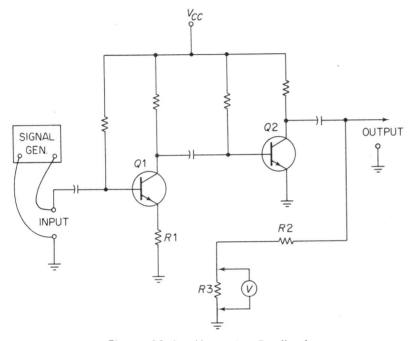

Figure 11-4. Measuring Feedback

The voltage gain in decibels at any given frequency is the ratio

$$\mathrm{dB} = 20 \log \frac{V_{in}}{V_{R3}}$$

This ratio should be observed at all frequencies, and enough plots should be made to establish a complete frequency-response curve. The highest peak on the curve is designated as zero dB, and all other points of the curve will be in decibels below this zero reference. The plot may be similar to any bandpass frequency-response curve but will have some of its own characteristics.

When the response curve is plotted, the input frequency should be set to the value which produced the highest point on the curve. The voltmeter is now placed across the output of $Q2$. The total amplitude of the output is measured and recorded. Call this voltage *output A*. Now remove $R3$ from the circuit and connect $R2$ back to the emitter of $Q1$. The circuit is now the same as the original circuit in Figure 11-3. With the same signal input, again measure the total output; this is *output B*. The voltage feedback in terms of decibels is

$$dB = 20 \log \frac{\text{output A}}{\text{output B}}$$

Use the same response curve previously plotted and draw a horizontal line which represents the zero-dB feedback level. The result is probably similar to that in Figure 11-5. The bandpass without feedback is from frequency A to B. These two points are 3 dB down from the two corner frequencies. With feedback the corner frequencies become C and D. These two points mark the intersection of the frequency curve and the zero-dB feedback. The rate of rolloff at points C and D reveals the stability of the amplifier. If this rolloff, in both cases, is less than 12 dB per octave, the circuit is stable. If the rolloff is 12 dB per octave, or greater, the amplifier will oscillate. The bandpass with feedback extends from 3 dB below point C to 3 dB below point D.

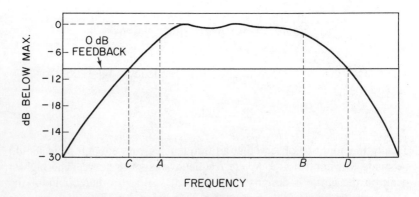

Figure 11-5. Frequency Response and Stability

## 11.2 Control Circuits

In the area of audio-frequency amplification, there are two types of feedback-dependent control circuits. These are gain control and tone control. The gain control alters the gain of one or more stages by varying the current feedback in an AF amplifier. The tone control varies the frequency response of an AF amplifier. The tone control is divided into two different circuits: *bass* control and *treble* control. The bass control varies the low-frequency response of an amplifier. The treble control varies the high-frequency response.

### 11.2.1 Gain Control

The gain of an AF amplifier is dependent on all the key resistors in the circuit. The gain can be varied by altering the resistance in collector, base, or emitter. It is not desirable to alter the collector resistance because this would alter the output impedance. Changing the base resistance would change the input impedance, and this feature is also undesirable. That leaves only the emitter resistance, and this has to be a resistor without a bypass capacitor. A practical example is illustrated in Figure 11-6.

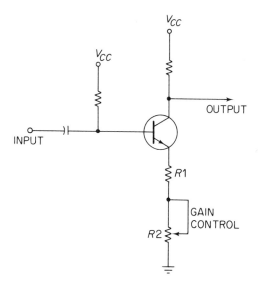

Figure 11-6. Practical Gain Control

The gain control $R2$ exerts a direct control on the quantity of degenerative current feedback. The series resistor $R1$ is most important when $R2$ is set to a very low resistive value. Without $R1$, there is zero resistance in the emitter circuit. This condition would result in zero feedback, strong instability, and possibly thermal runaway. Setting the value of $R1$ to one twentieth of the collector resistance gives the stage a maximum gain of 20. The maximum gain then decreases as the value of $R2$ increases.

The variable resistor $R2$ should be a noninductive, composition-type potentiometer with a wattage rating of about 2 W. The total resistance of $R2$ is decided by the quantity of control desired. If it is desired to decrease the gain all the way down to unity, then $R1 + R2$ should be equal to $R_C$. For example: with an $R_C$ of 6000 $\Omega$, $R1$ of 300 $\Omega$, and $R2$ of 5700 $\Omega$, the gain control can vary the gain from a maximum of 20 all the way down to unity.

### 11.2.2   Treble Control

A treble control provides a means of controlling the high-frequency response of an AF amplifier. This is usually accomplished by varying the level of a frequency-dependent voltage feedback. Two stages are generally involved, with a treble control as illustrated in Figure 11-7.

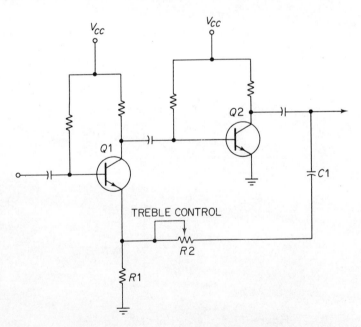

Figure 11-7.   Practical Treble Control

Transistor $Q1$ has current feedback independent of the treble control. The feedback voltage divider consisting of $R1$, $R2$, and $C1$ adds a frequency-dependent voltage feedback to the circuit. Resistor $R2$ is the treble control and it varies the voltage feedback by changing the resistance ratio between $R1$ and $R2$. There will be voltage feedback at all frequencies, but the higher the frequency, the larger the feedback. This change in feedback with respect to frequency is caused by the fact that the $X_c$ of $C1$ decreases as frequency increases.

The final values for $C1$ and $R2$ are best determined by experimentation. A good trial value for $C1$ is

$$C1 = \frac{f \times R1}{125}$$

where $C1$ is capacitance in farads, $f$ is the high-frequency limit in hertz, and $R1$ is in ohms.

A trial value for $R2$ should be based on the maximum expected gain times $R1$. If $R1$ is 50 Ω and the expected gain is 40, then

$$R2 = 40 \times 50$$
$$= 2 \text{ k}\Omega$$

Another arrangement for a treble-control circuit is illustrated in Figure 11-8.

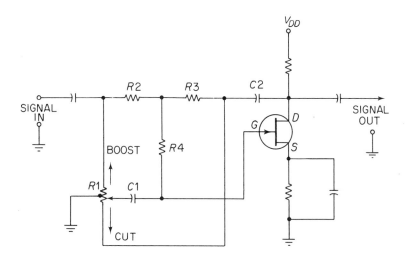

Figure 11-8. JFET Treble Control

This circuit is a JFET operational amplifier with a voltage-feedback treble control. The signal path for the input is through R2 and R4 in parallel with the top half of R1. Another signal path is R2 in series with R3, and this combination in parallel with the bottom half of R1. This same network forms the feedback circuit.

Resistor R1 is the treble control. It is a linear control potentiometer with a grounded center tap. The variable contact on R1 moves equidistance in both directions from ground. Moving the control toward the top of R1 taps off more signal and less feedback. Since the feedback is degenerative in nature, a decreasing feedback, or an increasing signal, is a boost to the gain of the amplifier. Moving the control downward on R1 reduces the signal and increases the feedback. A reduction in signal, or an increase in feedback, is a cut to the gain of the amplifier.

Trial values for the critical components should be based on the frequency and the expected gain. The degree of control is also an important consideration. For an average AF amplifier with control from 400 Hz to 10 kHz, the following values may suffice:

$$R1 = 500 \text{ k}\Omega$$
$$R2 = 100 \text{ k}\Omega$$
$$R3 = 100 \text{ k}\Omega$$
$$R4 = 500 \text{ k}\Omega$$
$$C1 = 100 \text{ pF}$$

With the component values indicated, the approximate extent of the boost and cut controls are illustrated in Figure 11-9. This curve repre-

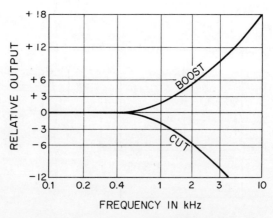

Figure 11-9. Extent of Treble Control

sents the two maximum extremes of the control setting. Intermediate settings of the control will provide smaller levels of control for either boost or cut. The setting of the control has no effect on the low frequencies or the center frequencies. The capacitors in the circuit are planned to allow effective feedback only after the signal reaches the minimum frequency where control is desired. In this case, the control becomes minimally effective at about 400 Hz. For intermediate settings of the control, the corner-frequency rolloff shifts to the right to a higher frequency than that indicated on the graph.

The capacitance of $C1$ is matched to the resistance of $R1$. This capacitor should present an $X_C$ that is as large as practical when compared to the resistance of $R1$ at the highest frequency to be boosted. In this circuit, the highest frequency is 10 kHz. At this frequency, the 100-pF capacitor has an $X_C$ of

$$X_C = \frac{1}{2\pi f C}$$
$$= \frac{0.159}{10 \times 10^3 \times 100 \times 10^{-12}}$$
$$= \frac{0.159 \times 10^6}{1}$$
$$= 159 \text{ k}\Omega$$

This is approximately one third of the 500-k$\Omega$ resistance of $R1$. A 100-pF capacitor was chosen for $C1$ on the premise that it would present a very small load for the input circuit and still be large enough to be immune to the stray capacitance. The resistance of $R1$ was then selected at a ratio of about 3:1.

The values of $C1$ and $R4$ both have an effect on the degree of action produced by the treble control. Increasing either of these values will provide more treble control. The values of $R1$ and $R3$ also have some effect. Increasing the resistance of $R1$ increases the amount of treble boost. Increasing the resistance of $R3$ increases the amount of treble cut.

### 11.2.3 Bass Control

A slight modification on the feedback circuit of Figure 11-8 will change the control from treble to bass. The bass control will exert either a boost or a cut for low-frequency signals and have no effect on middle or high frequencies. Figure 11-10 is the JFET operational amplifier with bass control. Resistors $R2$ and $R3$ are the same resistors used with the treble control. Their value was set at 100 k$\Omega$ each in the

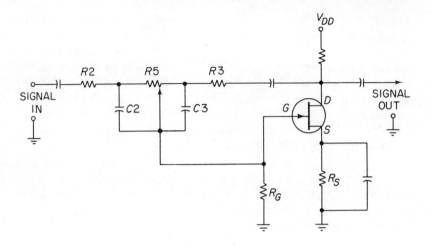

Figure 11-10. JFET Bass Control

previous circuit. These values are as good as any for the bass control; the important factor here is that the two resistors be of equal value. Resistor $R5$ is the bass control, and it should be a linear potentiometer with a total resistance of about 1MΩ. The resistance of $R5$ should be about 10 times that of either $R2$ or $R3$. It is desirable to have the value of $R5$ as large as practical. So if its value is more than 1 MΩ, the resistance of $R2$ and $R3$ should be increased accordingly. For a corner frequency of about 300 Hz, the capacitance for $C2$ and $C3$ should each be 5000 pF.

Moving the arm of $R5$ to the left increases the signal and reduces the feedback. This action provides a boost to the low-frequency signals. Moving the arm of $R5$ to the right decreases the signal and increases the feedback. This action provides a cut to the low-frequency signals. With the trial values indicated, the approximate extent of the boost and cut controls is illustrated in Figure 11-11.

These frequency-response curves represent the two extreme settings of $R5$. Intermediate settings of the control will provide smaller levels of control for either boost or cut. The setting of the control has no effect on the center frequencies. For intermediate settings of the control, the corner-frequency rolloff shifts to the left to a lower frequency than that indicated on the graph. A stronger low-frequency boost can be obtained by increasing the value of $C3$ and $R3$. A sharper cut can be obtained by increasing the value of $C2$ and $R2$.

Audio Feedback

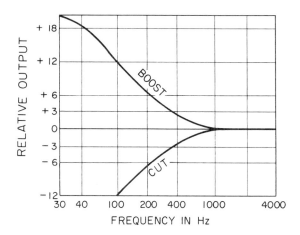

Figure 11-11. Extent of Bass Control

### 11.2.4 Complete Tone Control

A careful combining of the circuits in Figures 11-8 and 11-10 would result in a complete, practical, feedback tone control. In fact, it would produce the circuit in Figure 11-12. This is a somewhat simplified version of the Baxendall tone-control circuit that was named for its designer. The Baxendall complete tone-control circuit incorporates all

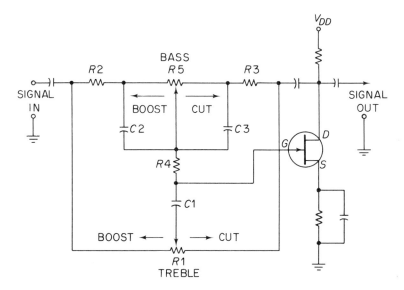

Figure 11-12. Baxendall Tone-Control Circuit

the action previously discussed under both treble and bass. The same component identifiers are used in all three schematics for ease of comparison. The bass and treble controls act as two independent circuits, and there is no detrimental interaction. The overall frequency response of this tone-control circuit is a combination of the graphs in Figures 11-9 and 11-11.

## 11.3 Results of Feedback

Feedback of some type is present in all electronic circuits regardless of the intent of the designer. Therefore, all audio-frequency amplifiers are feedback amplifiers. The designer can substantially increase the performance of his amplifier section by inserting appropriate feedback loops.

### 11.3.1 Broader Bandwidth

Degenerative feedback as discussed in this chapter produces a wider frequency-response curve in all cases. It also improves the flat magnitude or linear phase response.

### 11.3.2 Control over Impedance

The input and output impedances are altered by degenerative feedback. These impedance factors can be controlled, increased or decreased, by applying the proper type of feedback.

### 11.3.3 Reduction of Sensitivity

Oversensitivity is a fault in AF amplifiers. This fault is automatically overcome by the same feedback arrangement that broadens the bandwidth.

### 11.3.4 Reduction of Distortion

Since high fidelity is the ideal goal of all audio-frequency circuit designers, any type of distortion is detrimental. With large input signals, nonlinear distortion appears in the output of the amplifier. This distortion is produced by the nonlinear characteristics of the amplifying device. Feedback can be used to reduce this nonlinear distortion and to greatly increase the fidelity of the stage. It should be noted that distortion in the input signal cannot be helped by feedback.

### 11.3.5 Reduction of Gain

Sometimes it is desirable to reduce the gain of an amplification stage, but generally this is *not* true. In most cases, the loss of gain is the price that must be paid to enjoy the other advantages of feedback. In all cases, degenerative feedback reduces the gain of the amplifier. This loss of gain is not a serious matter. If need be, a few more stages of amplification can be inserted to compensate for the loss.

SECTION FOUR

# RADIO-FREQUENCY CIRCUITS

Radio-frequency (RF) circuits may be designed for any frequency from the top of the AF band up to the visible light frequencies. This is a wide band, and no single amplifier is capable of handling the entire band. With the exception of video circuits, which loosely fall into this category, RF circuits are tuned to amplify a narrow band of frequencies at a selected portion of the radio-frequency spectrum. This radio-frequency spectrum ranges from about 3 kHz to roughly 300 GHz.

CHAPTER TWELVE

# Radio-Frequency Amplifiers

*Radio-frequency amplifiers* are grouped into preamplifiers, intermediate-frequency (IF) amplifiers, power amplifiers, and video amplifiers. The particular name of each group is descriptive of the function that the group performs. All RF amplifiers are tuned circuits relying heavily on *LC* resonant circuits for frequency selection. With the exception of the video amplifier, they are narrow-band circuits.

## 12.1 Resonant Circuits

Any combination of inductance and capacitance is a resonant circuit at some frequency. At the resonant frequency, the capacitive reactance ($X_C$) is equal to the inductive reactance ($X_L$). When connected as a parallel tank circuit, the resonant circuit becomes a very high impedance to the resonant frequency. It provides maximum amplification for the resonant frequency and a narrow band of frequencies extending to either side of the resonant frequency. Therefore, it is a bandpass filter when the tank circuit is connected as a load impedance or in parallel with the signal path. In series with the signal path, the parallel tank circuit becomes a band reject filter.

When capacitor and inductor are in series, they form a series resonant circuit. The series resonant circuit offers a zero impedance to the resonant frequency. When connected in series with the signal path, the series resonant circuit is a bandpass filter; it passes the resonant frequency and a narrow band of frequencies to either side of the resonant frequency. When the series resonant circuit is connected in parallel to the signal path, it becomes a band reject filter. It shorts out the resonant frequency and the narrow band to either side of the resonant frequency.

### 12.1.1 Series Resonance Characteristics

The relationships among capacitance, inductance, reactance, and frequency in resonant circuits are important to the design of *RF* circuits. These relationships for a series resonant circuit are expressed in Figure 12-1 in the form of equations. The equations are based on the resonant frequency. Equation (a) states that the resonant frequency is equal to $1/2\pi$ multiplied by the square root of the inductance times the capacitance. Equations (b) and (c), simply put, indicate that at resonance the inductive and capacitive reactances are equal to each other. Since the two reactances are in series and are equal, they completely cancel one another. This means that at resonance, the impedance is zero as expressed by equation (d).

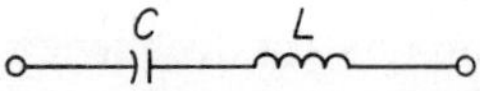

SERIES RESONANT CIRCUIT

(a) $f = \dfrac{1}{2\pi\sqrt{LC}}$    (c) $2\pi fL = \dfrac{1}{2\pi fc}$    (e) $Q = \dfrac{X_L}{R} = \dfrac{X_C}{R}$

(b) $X_L = X_C$    (d) $Z = 0$    (f) $BW = \dfrac{f}{Q}$

Figure 12-1. Characteristics of Series Resonant Circuits

Since the circuit must always have a finite amount of resistance, the $Q$, or quality, of the circuit is affected by this resistance. The $Q$ is equal to either $X_L$ or $X_C$ divided by the circuit resistance. This relationship is expressed in equation (e). The bandwidth is inversely related to $Q$, as indicated by equation (f). The bandwidth is equal to the resonant frequency divided by the $Q$ of the circuit.

### 12.1.2 Parallel Resonance Characteristics

The parallel resonant tank circuit is most often seen in RF amplifiers. These tuned tank circuits are used for output load impedances and for input coupling. The critical relationships are expressed as equations in Figure 12-2.

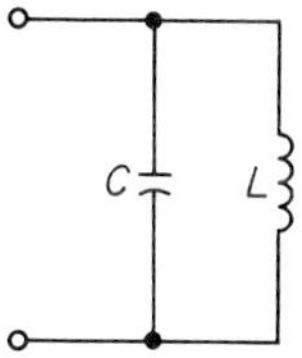

# Radio-Frequency Amplifiers

PARALLEL RESONANT TANK

(a) $f = \dfrac{1}{2\pi\sqrt{LC}}$

(b) $X_L = X_C$

(c) $2\pi fL = \dfrac{1}{2\pi fC}$

(d) $Z = \dfrac{(X_L)^2}{R_L} = \dfrac{(X_C)^2}{R_L}$

(e) $Q = \dfrac{I_C}{I_t} = \dfrac{X_L}{R_L} = \dfrac{Z}{X_C}$

(f) $BW = \dfrac{f}{Q}$

Figure 12-2. Characteristics of Parallel Tank Circuits

The conditions of resonance are the same in series and parallel tank circuits: equations (a), (b), and (c). It should be remembered that the conditions of resonance, as expressed here, are true only for high-Q circuits. A circuit is considered to be high $Q$ when the $Q$ is equal to 10 or more.

In equation (d), both $X_L$ and $X_C$ are extremely high values. Since $X_L$ and $X_C$ are equal, the impedance is the square of either reactance divided by $R_L$. The $R_L$, in this case, is the resistance of the inductor. The $Q$ of the circuit can be determined in several ways, as indicated in equation (e). Current $I_t$ is the line current, and current $I_C$ is the capacitor current. The other quantities in this equation are self-explanatory. The bandwidth, equation (f), is the same as for series resonant circuits: frequency of resonance divided by the $Q$ of the tank.

The impedance of a parallel tank circuit can also be calculated directly from the values of capacitance, inductance, and resistance. Again, this produces true results only at the resonant frequency:

$$Z = \dfrac{I}{CR_L}$$

The impedance of a parallel tank circuit is very high at resonance and drops off sharply for frequencies above and below resonance. The graph in Figure 12-3 illustrates the impedance change in response to changes in frequency for both low- and high-Q circuits.

The width of the impedance–frequency curve is inversely proportional to $Q$:

$$Q = \dfrac{Z}{X_C}$$

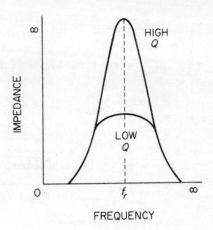

Figure 12-3.  Impedance–Frequency Curve

### 12.1.3 Selectivity and Bandwidth

The amplitude of the signal that is developed across a tank circuit is directly proportional to $Q$ while the bandwidth is inversely proportional to $Q$. The relationships of amplitude, frequency, $Q$, and bandwidth are illustrated in Figure 12-4.

Selectivity is a measure of how well a circuit accepts one frequency and rejects all others. This means that selectivity decreases as bandwidth increases. A high-$Q$ circuit has a narrow bandwidth and a high degree of selectivity. A low-$Q$ circuit has a wide bandwidth and poor selectivity.

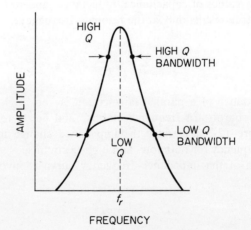

Figure 12-4.  Amplitude–Frequency Curve

# Radio-Frequency Amplifiers

The bandwidth is defined as the frequency difference between the low half-power point and the high half-power point on the amplitude–frequency curve. These points are indicated by dots on the curves in Figure 12-4. The bandpass is the group of frequencies contained in the bandwidth. For example, with a resonant frequency of 10 MHz and a bandwidth of 2 MHz, the bandpass is 9 to 11 MHz.

### 12.1.4 Effects of Loading

Any resistor connected in parallel with a tank circuit places a load on the circuit with some rather drastic effects on $Q$, amplitude, and bandwidth. Consider the circuit in Figure 12-5. Without the 10-k$\Omega$ parallel resistor, this circuit has a $Q$ of 95.13, a resonant frequency of 500 kHz, and a bandwidth of 5.3 kHz. At resonance, $X_L$ and $X_C$ are each equal to 637.4 $\Omega$.

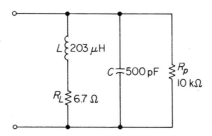

Figure 12-5.  Parallel Loading

When the 10-k$\Omega$ parallel resistor is added, many things happen. We can determine the capacitor current by assuming a voltage. For convenience, use 10 V:

$$I_C = \frac{E_a}{X_C}$$

$$= \frac{10 \text{ V}}{637.4 \ \Omega}$$

$$= 15.7 \text{ mA}$$

The current in the line is equal to the same 10 V divided by the parallel resistance:

$$I_t = \frac{E_a}{R_P}$$

$$= \frac{10 \text{ V}}{10 \text{ k}\Omega}$$

$$= 1 \text{ mA}$$

The $Q$ is equal to $I_C$ divided by the line current:

$$Q = \frac{I_C}{I_t}$$

$$= \frac{15.7 \text{ mA}}{1 \text{ mA}}$$

$$= 15.7$$

So the $Q$ of the circuit dropped from 95.13 to 15.7 because of the parallel loading. The bandwidth also changes drastically:

$$\text{BW} = \frac{f_r}{Q}$$

$$= \frac{500 \text{ kHz}}{15.7}$$

$$= 31.8 \text{ kHz}$$

The bandwidth has increased from 5.3 kHz to 31.8 kHz, with a corresponding reduction in selectivity.

In most cases, there is a resistance in parallel with a tank circuit. This resistance may be a physical resistor or the resistance of a transistor. In either case, the resistance must be considered because it drastically alters the characteristics of the tank circuit.

## 12.2 Radio-Frequency Power Amplifiers

Radio-frequency power amplifiers are used almost exclusively in radio and television transmitting equipment. The final stage in a transmitter must be a power amplifier. This final power amplifier produces the power to drive the antenna. Of course, there may be several stages of RF amplification, but generally most of the power amplification is provided by the last stage before the antenna.

There are several factors to consider in the design of an RF power amplifier. The transistor must be able to handle the current, and heat sinks are probably required to dissipate the excessive heat. The dc power supply must be isolated from the RF signal. A resonant network must be provided for signal amplification. This resonant network must match the output impedance of the transistor to the impedance of the line that connects the RF from the amplifier to the antenna. If the frequency is high, the leads must be kept as short as possible because a section of wire can take on the characteristics of an inductor.

### 12.2.1 Sample Circuit

There are a great many circuit configurations to choose from in RF power amplifier design. One possible layout is illustrated in Figure 12-6.

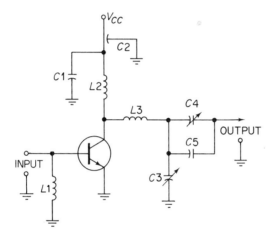

Figure 12-6. Sample Circuit of an RF Power Amplifier

Inductors $L1$ and $L2$ are radio-frequency choke coils. Inductor $L1$ acts as a high-impedance input load for developing maximum input to the transistor; it also isolates RF from ground. Inductor $L2$ offers a high impedance to the RF signal and helps to isolate it from the dc power supply. Capacitors $C1$ and $C2$ are RF bypass capacitors. Capacitor $C1$ is a normal bypass capacitor, while $C2$ is a feed-through bypass. These two capacitors are in parallel, and the total capacitance should provide a near short to the frequency used.

$$C_t = C1 + C2$$

$$X_c = \frac{1}{2\pi f C}$$

The value of the RF choke coils is determined by a trade-off between inductance and current-handling ability. The minimum current-handling ability should exceed the expected maximum current by at least 10 percent. The inductive reactance is dependent upon frequency. Each RF choke should produce an $X_L$ between 1000 and 3000 Ω at the operating frequency:

$$X_L = 2\pi f L$$

The values given for RF choke inductors and bypass capacitors are intended as trial values. In the final analysis, these components must eliminate all detectable RF components from the dc power supply. If RF signals are present in the dc line, the value of bypass capacitance or RF choke inductance must be increased. This increase must be handled with care. Too much bypass capacitance will cause feedback and oscillations. Too much inductance can cause a decrease in gain and efficiency.

The resonant circuit is formed by $L3$ and $C3$. The variable capacitor, $C3$, can be tuned to vary the frequency. The desired operating frequency should be the center of the tuning range. The values of $L3$ and center capacitance for $C3$ are based on the operating frequency. The operating frequency is the resonant frequency of this circuit:

$$f_r = \frac{1}{2\pi\sqrt{LC}}$$

$$X_L = 2\pi f L$$

$$X_C = \frac{1}{2\pi f C}$$

$$\alpha = X_L$$

Capacitor $C4$ is a loading capacitor. Its purpose is to match the output impedance of the transistor to the impedance of the coaxial line going to the antenna. Capacitor $C5$ is a fixed capacitor in parallel with $C4$. Using a fixed capacitor in this fashion reduces the required capacitance of $C4$ and enables the use of two small capacitors instead of one large capacitor. Capacitor $C5$ is also a safety factor; when $C4$ is set to a minimum value, a fixed minimum capacitance is still present. The fixed minimum capacitance prevents a severe mismatch of impedance, which could destroy the transistor.

A transistor connected in this manner will provide class C operation, and no special consideration of bias is necessary. If the common-emitter configuration is used, the emitter must be connected directly to ground, as illustrated. Any type of impedance between emitter and ground will cause degenerative feedback of an undesirable nature.

# Radio-Frequency Amplifiers

### 12.2.2 Selecting the Transistor

Information from the data sheets must be consulted for a suitable transistor. The transistor must be able to handle the current load produced when the full value of the supply voltage is on the collector. Select a transistor capable of handling the expected output power plus about 30 percent. The RF power amplifier is only about 70 percent efficient, so

$$P_{total} = \frac{P_{out}}{0.7}$$

The total collector current can be estimated by dividing the total output power by the total applied voltage:

$$I_C = \frac{P_{total}}{V_{CC}}$$

Make sure that power and current ratings exceed the calculated values by a safe margin. And be sure to consider the power derating factor. Now consider the frequency of operation. Can the transistor deliver the required power at this frequency? Finally, consider the power gain and the expected input power to this stage. Is the gain sufficient to boost the input power to the required total power?

### 12.2.3 Planning the Resonant Circuit

The resonant network in Figure 12-6 is composed of $L3$ and $C3$. The value of $L3$ and $C3$ should be chosen so that they form a resonant circuit for the operating frequency when $C3$ is set near its midpoint.

The input impedance of the resonant circuit must match the output impedance of the transistor, and the output impedance of the resonant circuit must match the input impedance of the line to the antenna. The input impedance to the line is easily determined. This is probably a coaxial line with a characteristic impedance of 70 Ω. The output impedance of the transistor is a bit more complex. The transistor has an output resistance $r_p$ and an output capacitance $C_o$. The output capacitance is given on a graph as part of the furnished data. The output resistance is easily calculated using this equation:

$$r_p \approx \frac{(V_{CC})^2}{2 \times P_{total}}$$

This is an approximation, but it is close enough for average design.

The output capacitance, $C_o$, varies with frequency. So a frequency-dependent graph similar to that in Figure 12-7 is essential.

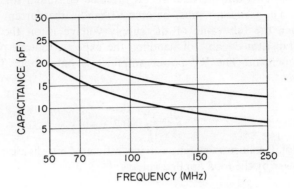

Figure 12-7. Output Capacitance

The two lines give a tolerance of about 5 pF for any operating frequency. The center point between the two lines is the best estimate. For example, on this graph, the output capacitance for a frequency of 100 MHz is about 15 pF.

## 12.3 Radio-Frequency Preamplifiers

The RF *preamplifier* is one of several RF voltage amplifiers. Some of the other voltage amplifiers are IF amplifiers, limiters, and video amplifiers. These circuits are found in communications receivers and in instruments with receiver-type functions. Most RF voltage amplifiers have a tuned input and a tuned output. They can be tuned to pass a narrow band of frequencies, as is the case with IF amplifiers, or a broad band of frequencies, as in the video amplifier.

The RF preamplifier is usually placed as the first stage in a telecommunications receiver circuit. Radio, radar, and television are examples of equipment utilizing telecommunications-type receivers. The primary purpose of such a stage is to improve the signal-to-noise ratio of the received signal. The objective is to amplify the very weak signal from the antenna before it has a chance to mix with locally generated noise. Since the incoming signal couples directly from the antenna into the preamplifier, the preamplifier input is tunable over a wide frequency range. This wide tuning range enables the selection of any signal that falls within its tunable range. In the case of commercial radio and television receivers, the tunable range includes the entire broadcast band.

### 12.3.1 Sample RF Preamplifier Circuit

The schematic layout in Figure 12-8 will serve as a starting point for the design of an RF preamplifier. Transformers $T1$ and $T2$ are tuned RF transformers. This stage is concerned only with the secondary of $T1$ and the primary of $T2$, but the primary of $T1$ and the secondary of $T2$ may also be tuned tank circuits. The secondary of $T1$ and $C1$ form a resonant tank circuit. The primary of $T2$ and $C5$ form another resonant tank circuit. The station selector control varies $C1$ and $C5$ so that these two tanks are always tuned to the same frequency. In fact, the two capacitors should be two parts of a ganged tuning capacitor. The station selector also controls a third capacitor in the local oscillator of the mixer stage.

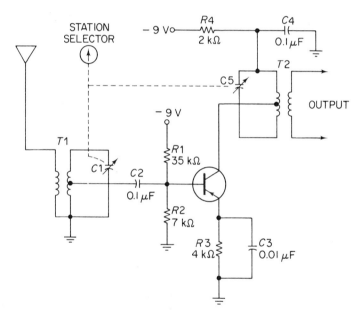

Figure 12-8. Sample RF Preamplifier Circuit

Capacitor $C2$ is a coupling capacitor. The RC coupling from the $T1$ tank circuit is formed by $C2$ and $R2$. Bias for the transistor is provided by the $R1R2$ voltage divider and $R3$.

Stability swamping is provided by $R3$ while the bypass capacitor $C3$ prevents degeneration across this resistor. Decoupling of the RF signal is provided by $R4C4$. Resistor $R4$ may be replaced by an RF choke coil.

### 12.3.2 Selecting Components

Voltage gain for the stage is calculated from the base of the transistor to the secondary of $T2$. This gain is dependent upon the amplification factor of the transistor and the characteristics of $T2$. A gain between 10 and 20 is satisfactory for this RF preamplifier. As a rough estimate, the stage gain can be calculated as follows:

$$A_V = \beta \times \frac{1}{Z_r} \quad \text{where } Z_r = \sqrt{\frac{Z_p}{Z_s}} = \frac{N_p}{N_s}$$

Most transistors intended for voltage amplification will readily fill the bill for IF amplification. The only problem could come from frequency-handling capability. Make sure that it can handle the entire band of frequencies intended for this receiver. The power level here is very low, so the temperature is *not* a problem.

Coupling capacitor $C2$ should offer a low opposition to the lowest operating frequency. For an AM broadcast receiver, the lowest frequency is 500 kHz. The 0.01-$\mu$F capacitor used here offers about 32 $\Omega$ of reactance to the frequency. The same is true for capacitors $C3$ and $C4$. However, $C3$ and $C4$ may be of slightly smaller values.

The values of $R1$, $R2$, and $R3$ must be chosen so as to provide a base-to-emitter differential that sets the proper operating point. The resistance of $R4$ is *not* critical, but it should be relatively high in respect to the reactance of $C4$. If the designer so desires, an RF choke coil can be used instead of $R4$.

The level of voltage used for the supply is not critical so long as the limits of the transistor are not exceeded. The minimum voltage level should be about four times the expected amplitude of the output voltage. The output of this first stage is generally not more than 2 V; so a 9-V supply is adequate.

The RF transformers are off-the-shelf items, and it may be necessary to design the stage around the transformer characteristics. Transformers of this type are generally rated as to primary and secondary impedance and current capacity. Since a tunable capacitor is connected across the secondary of $T1$ and the primary of $T2$, a "loopstick" transformer should be used in both cases. A standard tuning capacitor with a loopstick transformer will provide tuning across the entire broadcast band. When the tuning capacitor is set at midrange, the two tank circuits should be resonant at the midband frequency.

## 12.4 IF Amplifiers

The *IF amplifier* falls into a group of tuned, narrow-band amplifiers. It is a voltage amplifier similar in many ways to the RF preamplifier. It generally has a tuned tank input and output. This type of amplifier is sometimes called a bandpass amplifier because it passes a particular center frequency with a narrow band. The band extends slightly to either side of the center frequency.

### 12.4.1 Sample IF Amplifier Circuit

An IF amplifier for an AM radio is generally tuned to a center frequency of 455 kHz. It passes a band of frequencies from about 450 kHz to about 460 kHz. A sample circuit for such an amplifier is illustrated in Figure 12-9. The tuning elements in this circuit are sealed-unit fixed-capacitor slug-tuned IF transformers. These transformers may be shielded to prevent noise pickup. A tuned output from the previous stage and a tuned input for this amplifier are provided by $T1$. A tuned

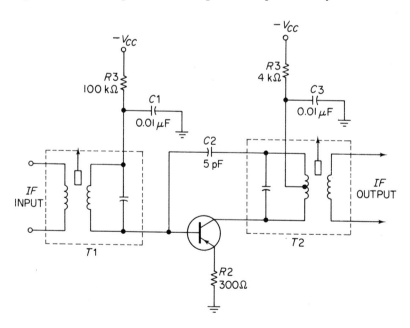

Figure 12-9. Sample IF Amplifier Circuit

output from this stage and a tuned input for the next stage are provided by $T2$.

The dc power supply must be protected from the IF signals. This protection is provided by two decoupling networks: $R1C1$ and $R3C3$. Resistor $R1$ is also part of the biasing network since the base current of the transistor must pass through it.

Resistor $R2$ works with $R1$ in providing dc bias. This resistor ($R2$) has two additional functions. It provides thermal stabilization and degenerative feedback. The degenerative feedback squelches any tendency of the amplifier toward oscillation.

Feedback is also provided by $C2$. This feedback is intended to neutralize the internal feedback of the transistor.

### 12.4.2 Selecting Components

The IF transformers can be obtained as off-the-shelf items. They are usually constructed with slug-tuned primaries and low-resistance secondaries. The average secondary impedance is about 800 Ω. This arrangement makes it easy to obtain an optimum impedance match and still have the benefit of resonant tank circuits.

Both transformers may be tuned to the center frequency of the bandpass, 455 kHz, but this *is not* the best procedure. An IF amplifier seldom stands alone. Usually it is one of several amplifiers that compose an IF strip. The transformers are usually stagger-tuned, one slightly below 455 kHz, the next slightly above, the next below, and so on. This stagger tuning, if skillfully handled, will provide a uniform gain for all frequencies in the bandpass. These IF transformers are usually constructed with the tunable slug in the primary.

Each of the decoupling capacitors should offer a very low impedance to the center frequency. These 0.01-$\mu$F capacitors each have a reactance of about 35 Ω. The value of $R3$ is not critical, but it must be large in comparison to the reactance of $C3$. The value of $R1$ is dictated by the level of $V_{CC}$ and the quantity of base current. If $R1$ is supplying sufficient bias for the transistor, its value is plenty high for the job of decoupling.

The transistor should be selected on its ability to handle the frequency, produce the required gain, and generate a minimum of noise. The required gain is about 30 in most cases. Whatever the gain required, add about 25 percent for transformer losses. This is a low-power stage, so no special power precautions are required.

## 12.5 Video Amplifiers

A *video amplifier* is of necessity a wide-band amplifier because video signals are composed of a fundamental frequency and many harmonics of that frequency. The perfect square wave is composed of a fundamental frequency and an infinite number of odd harmonics. The stepped waveshape of television video signals contains frequencies that range from about 30 Hz to something in excess of 4 MHz. It would be nice to have a video amplifier with a uniform frequency response from 30 Hz to 4 MHz. That is a good goal to strive for, but practical circuit designers must compromise for much less.

A carefully designed *RC* coupling network can adequately handle frequencies from 100 Hz to 50 kHz. With the addition of boosting circuits, this response can be pushed below and above these limits. The use of *RC* coupling, along with both low- and high-frequency compensation, is the most practical approach to video amplifier design when using bipolar transistors.

### 12.5.1 Low-Frequency Problem

The *RC* response with 100 Hz as the low end is far from satisfactory. Two steps can be taken to extend the low-frequency end of the response curve. The circuit can be designed to provide less attenuation for the low frequencies and to amplify the low frequencies more than normal. Examine the circuit in Figure 12-10 and observe the reason for low-frequency attenuation.

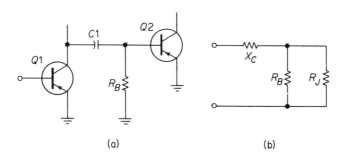

Figure 12-10. RC Coupling and an Equivalent Circuit

In Figure 12-10a, the signal couples across $C1$ and is developed across $R_B$. The equivalent circuit (12-10b) shows that $X_C$ is in series with the signal path. Any voltage drop across $C1$ is a direct reduction of the signal amplitude. Only one action can be taken to reduce this loss. Since $X_C$ is inversely proportional to $C$, the reactance can be lowered by increasing the capacitance:

$$X_C = \frac{1}{2\pi f C}$$

At a frequency of 50 Hz, a 0.1-$\mu$F capacitor offers a reactance of about 32,000 $\Omega$. A 1-$\mu$F capacitor at the same frequency has an $X_C$ of about 3200 $\Omega$. Unfortunately, as capacitance increases, the physical size of the capacitor also increases, and so does the cost. Let the limit be set by personal judgment. Make the capacitor as large as practical.

Once the signal has gotten past the capacitor, it is developed across $R_B$. But notice in the diagram that the junction resistance $R_J$ is in parallel with $R_B$. The junction resistance for the common emitter is on the order of 1500 $\Omega$; so the maximum possible resistance of the combination is about the same as the resistance of $R_J$. It is highly desirable to keep this combined resistance as high as possible. This is done by keeping the value of $R_B$ at least 10 times as large as the value of $R_J$. By keeping $X_C$ as low as possible and the $R_B R_J$ combination as high as possible, the attenuation of the input low-frequency signal is reduced to a minimum.

Next, the circuit should be modified to provide more amplification for the lower frequencies. This can be done by adding an $RC$ compensation circuit, as indicated in Figure 12-11. The network composed of $R2$ and $C2$ has been added in series with the collector load for $Q1$. At

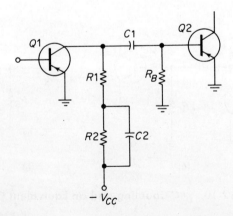

Figure 12-11. Low-Frequency Compensation

# Radio-Frequency Amplifiers

high and middle frequencies, C2 is a short, and it removes R2 from the circuit. As frequencies get lower, $X_C$ increases and places more impedance in series with R1. As the collector load impedance increases, amplification increases. When $X_C$ becomes 10 times the value of R2, the load impedance is maximum and is equal to $R1 + R2$.

The desired low-frequency end of the response curve is 30 Hz. The values of R2 and C2 should be chosen on this premise. At 30 Hz, $X_C$ must be at least 10 times the resistance of R2.

### 12.5.2 High-Frequency Problem

At the high-frequency end of the response curve, the enemy takes the form of distributive capacitance. This distributive capacitance appears in parallel with the signal path, and at high frequencies, the $X_C$ becomes small and tends to short out the signal. The distributive capacitance is illustrated in Figure 12-12.

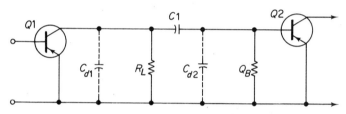

Figure 12-12. Distributive Capacitance

Here the distributive capacitance is divided into $C_{d1}$ and $C_{d2}$ to represent the output capacitance of one stage and the input capacitance of the other. A distributive capacitance is chiefly composed of two ingredients: the internal capacitance of the transistor and capacitance between the external leads. Transistors must be selected with a minimum of internal capacitance, and leads must be kept widely separated and as short as possible. These three precautions will keep the distributive capacitance to a minimum value.

A minimum of distributive capacitance reduces the shunting effect to a minimum, but that is not enough. More amplification must also be provided for the high frequencies. The high frequencies can be boosted by shunt-peaking and by series-peaking circuits. Both shunt and series peaking should be used for a video amplifier. These circuits are illustrated in Figure 12-13. Shunt peaking is provided by L1 and $C_{d1}$, and series peaking is provided by L2 and $C_{d2}$. The value of L1 should be selected to form a resonant circuit with $C_{d1}$ at the top frequency. The values of L2 and C2 must also form a resonant circuit at the same or a slightly higher frequency. For example, if $C_{d1}$ is 10 pF, it has an $X_C$ of

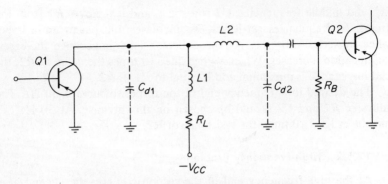

Figure 12-13.  Shunt and Series Peaking

3975 Ω at a frequency of 4 MHz. The $X_L$ of $L1$ must also be 3975 Ω at 4 MHz. Therefore, the inductance of $L1$ should be

$$L = \frac{X_L}{2\pi f}$$

$$= \frac{3975}{6.28 \times 4 \times 10^6}$$

$$= 158 \ \mu H$$

The $Q$ of the two tank circuits must be kept fairly low to prevent a ringing problem. Sudden changes of voltage in a high-$Q$ circuit causes overshoot and sets up transit oscillations on the output from the resonant circuit.

### 12.5.3  Sample Bipolar Video Amplifier

An $RC$ coupling network with both low-frequency and high-frequency compensation networks will serve as a trial circuit for a bipolar video amplifier. Such a circuit is illustrated in Figure 12-14. This circuit shows a combination of low-frequency compensation and both types of high-frequency compensation. The distributive capacitance is not shown, but $C_{d1}$ and $C_{d2}$ are still present, as previously described. The only new components in this circuit are the resistors in parallel with the two coils. These resistors are built-in loads to keep a low-$Q$ circuit and prevent ringing. The peaking coils can be purchased with built-in resistors, as indicated in the circuit.

As a starting condition it is suggested that $X_{C2}$ be 10 times the value of $R2$ at a frequency of 30 Hz. The value of $L1$ should form a resonant circuit with $C_{d1}$ at a frequency of 3 MHz. The value of $L2$ should form

# Radio-Frequency Amplifiers

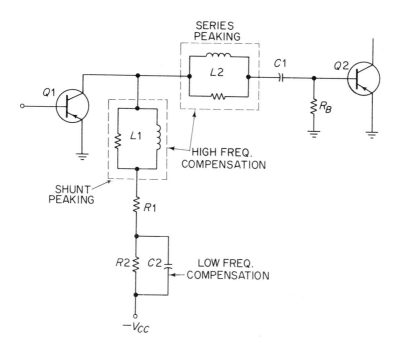

Figure 12-14.  Coupling for a Bipolar Video Amplifier

a resonant circuit with $C_{d2}$ at a frequency of 3.5 MHz. These are not the highest desired frequencies, but they are about the highest that can be expected.

### 12.5.4  Testing

Under test conditions, the designer should apply signals from 30 Hz to 4 MHz and plot the frequency response curve of the output from the collector of $Q2$. He should then experiment with slightly different values of $R2$, $C2$, $L1$, and $L2$ until the circuit produces a maximum bandwidth. The top of the response curve should be reasonably level, but it will have a few humps caused by resonant peaks.

### 12.5.5  JFET Video Amplifier

An easy way to solve the low-frequency problem in a video amplifier is to use a JFET instead of a bipolar transistor. The extremely high impedance of an $N$-channel JFET in a common-source configuration enables an $RC$ circuit with very little low-frequency attenuation. The input impedance is on the order of 10 M$\Omega$. So a very high resistance,

1 MΩ or more, can be used as an input resistor. This enables the use of a nominal-sized coupling capacitor. A 0.1-$\mu$F coupling capacitor and a 1-MΩ resistor will couple a 10-Hz signal with an attenuation of only 16 percent.

Using the JFET as the amplifying device only solves the low-frequency problem. The high-frequency compensation must still be included, as previously described.

CHAPTER THIRTEEN

# Modulation and Demodulation

*Modulation* is the process of impressing an intelligence on a carrier wave. *Demodulation* is the process of extracting the intelligence from a modulated carrier wave. *Modulation circuits,* such as RF modulators and frequency multipliers, are essential parts of a transmitter. *Demodulation circuits,* such as mixers, detectors, and discriminators, are essential parts of a receiver.

## 13.1 Amplitude Modulators

An *amplitude modulator* impresses the intelligence on the amplitude of the RF carrier wave. There is a choice between high-level modulation and low-level modulation. High-level modulation usually takes place in the final stage of RF amplification. Low-level modulation takes place in the first stage of RF amplification.

### 13.1.1 Amplifier Modulator

High-level modulation is performed by a circuit called an *amplifier modulator.* Actually, an amplifier modulator is a modified RF power amplifier. Since RF power amplifiers were discussed in Chapter 12, the discussion here concerns only the special features. When the gain of an RF amplifier is controlled by a modulating wave, the stage becomes an amplifier modulator. The input signal to an amplifier modulator is a continuous RF at a constant amplitude and frequency. While this signal is being amplified, the injected modulating signal varies the gain of the stage. The output is an RF carrier wave with the amplitude varying according to frequency and amplitude of the modulating wave.

The modulating wave is usually an AF signal from a microphone or some other type of transducer. This audio signal can be injected into the amplifier stage in either of three ways. It can be applied to the base, the emitter, or the collector. Base injection and emitter injection are methods used in low-level modulation. Both of these methods require a low-level RF signal into the amplifier. It is difficult to obtain a high degree of modulation with a low-level modulator. Also, with low-level modulation, the modulated carrier wave must be greatly amplified through additional stages before it can be transmitted.

Collector injection is a method of high-level modulation. A sample circuit using this method is illustrated in Figure 13-1. With the exception of $T3$, the stage is a standard RF power amplifier. Resistors $R1$ and $R2$ form a voltage divider for base bias. Resistor $R3$ is part of the biasing circuit and adds stability. Capacitor $C3$ acts as an RF bypass to prevent signal degeneration. Capacitors $C1$ and $C5$ are RF decoupling capacitors. The RF signal is applied through the tuned secondary of $T1$ to the base of the transistor. The RF signal is amplified through the transistor and developed in the tuned primary of $T2$.

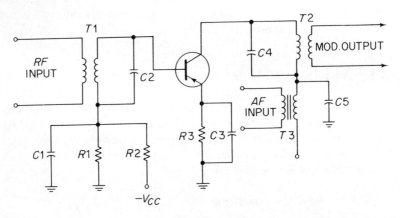

Figure 13-1. Collector Injection Amplifier Modulator

The AF modulating signal is applied to the primary of $T3$. The secondary of $T3$ is part of the collector load impedance. The impedance of this secondary varies at the audio rate of the modulating signal. The varying impedance of the $T3$ secondary varies the collector voltage on the transistor. The overall result is a transistor gain that varies with the audio signal. The RF signal in the tank circuit has an amplitude that varies at the audio rate. This amplitude-modulated wave is coupled from the secondary of $T2$ to the antenna or to another RF amplifier.

The voltage amplitude of the modulating signal at the secondary of $T3$ should match the amplitude of the RF signal in the primary of $T2$. Several stages of audio amplification may be required to bring the audio signal to the proper level. Transformer $T3$ is an audio transformer and may, or may not, have a tuned primary. The secondary of $T3$, as well as the primary of $T2$, must be capable of handling the current produced with full $V_{cc}$ on the collector.

### 13.1.2 Oscillator Modulator

An RF carrier wave can be amplitude-modulated while it is being generated. The stage that accomplishes this dual function is an *oscillator modulator*. It is a matter of personal preference whether the modulating signal is injected into the base, emitter, or collector of the transistor. The sample illustrated in Figure 13-2 uses base injection.

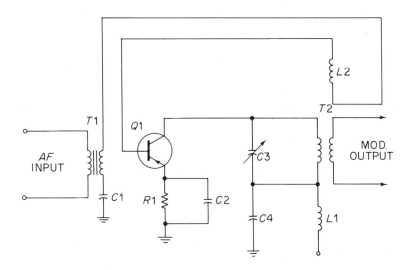

Figure 13-2. Base Injection Oscillator Modulator

Part of the biasing circuit has been omitted for simplicity. This circuit is a modified Armstrong oscillator. See Chapter Fourteen for a detailed discussion of several oscillators, including the Armstrong. The oscillator portion is covered here only briefly. The operating frequency is determined by the tank circuit in the collector of the transistor. Capacitor $C3$ is used to tune the tank to the desired frequency. Inductor $L2$ is a tickler coil. The tickler coil picks up a small portion of the RF from $T2$

and couples it back to the base of the transistor. This feedback is sufficient to sustain the RF oscillations in the tank circuit.

Capacitors $C1$ and $C4$ are decoupling capacitors. They must offer a low impedance to the RF signal. Capacitor $C1$ must be selected carefully to offer a very high impedance to the audio signal and a virtual short to the RF. The secondary of $T1$ and inductor $L1$ are RF chokes to improve the decoupling. Resistor $R1$, in the emitter circuit, is part of the biasing circuit, and it provides thermal stability. Capacitor $C2$ is an RF bypass to prevent RF signal degeneration.

The modulating audio signal input to the stage is the modification. The AF signal couples across $T1$. From the secondary of $T1$, the AF signal feeds through the tickler coil $L2$ to the base of the transistor. The audio signal varies the potential on the base of the transistor and controls the current and the gain of the transistor. The RF tank in the collector circuit oscillates at its set frequency, but the amplitude of these oscillations varies with the amplitude and frequency of the audio signal. The amplitude-modulated RF is coupled from the secondary of $T2$ to the antenna or to another RF amplifier.

Transformer $T1$ is an audio transformer. The audio signal on the primary is a reasonably large signal, but an ordinary AF transformer should suffice. An oscillator modulator of this type has a small degree of frequency modulation. This undesirable effect is caused from the collector-to-emitter capacitance inside the transistor. A shift in collector voltage causes a change in internal capacitance. The changing capacitance causes a slight change in the resonant frequency of the tank circuit. This shift in frequency of the tank causes frequency shifts in the carrier wave. This frequency modulation is undesirable, but the degree of modulation is low and can be tolerated.

## 13.2 Frequency Modulators

A *frequency modulator* impresses the intelligence into the *frequency* of an RF carrier wave. Most frequency modulators use some form of variable reactance to vary the frequency. Most frequency modulators are of the oscillator modulator type. The main differences between AM and FM oscillator modulators is the level of modulation and the degree of frequency modulation. In AM the modulation level is high, and the degree of frequency modulation is very low. In FM the modulation level is low, and the degree of frequency modulation is greatly enhanced.

### 13.2.1 Reactance Oscillator Modulator

The reactance oscillator modulator circuit is similar to the oscillator modulator previously described for amplitude modulation, but now the variable reactance is stressed instead of being incidental. A sample circuit is illustrated in Figure 13-3. The biasing, stabilizing, and decoupling circuits have been omitted.

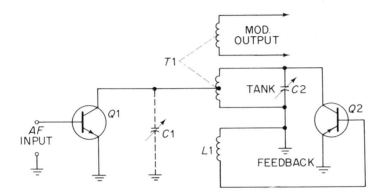

Figure 13-3. Sample Reactance Modulator

This circuit is a combination of an Armstrong oscillator and a variable-reactance transistor. The primary of $T1$ is the frequency-controlling tank circuit for $Q2$. Inductor $L1$ is the feedback loop for sustaining oscillation. The secondary of $T1$ couples the frequency-modulated output to the next stage. This is the oscillator portion of the circuit.

Transistor $Q1$ is a variable-reactance transistor. The AF modulating signal is coupled into the base of $Q1$. Capacitor $C1$ represents the variable capacitance between the collector and emitter of $Q1$. This capacitance varies with the bias on $Q1$, and the bias varies with the audio signal. Notice that $C1$ parallels a portion of the tank circuit. Any variation of $C1$ alters the resonant frequency of the tank circuit. As the resonant frequency changes, it causes the frequency of the RF signal to deviate from the center frequency. The result is an RF carrier with the frequency varying at an audio rate.

This reactance modulator is a low-level modulator. The RF signal is being generated in this stage and is a weak signal. The audio signal is also very weak. The modulated output must be further amplified before it can be transmitted. Usually several stages of amplification are required.

The reactance modulator produces small variations in the amplitude of the carrier wave. This is an undesirable feature and necessitates an additional stage to remove this amplitude modulation. The carrier wave is fed through a limiter stage which clips off all peaks that exceed a specified amplitude. The output from the clipper is a constant-amplitude frequency-modulated carrier wave.

In most cases, the output from the modulator is of a much lower frequency than the desired transmitter frequency. Frequency multipliers are then utilized to increase the carrier wave to the desired frequency.

### 13.2.2 Sidebands

In FM many sidebands are generated both above and below the center frequency of the carrier wave. All the intelligence is carried in these sidebands: the center frequency has no intelligence. All sidebands are significant until their amplitude decreases to almost zero. If the sideband amplitude is 1 percent of the amplitude of the carrier-wave center frequency, it is a significant sideband. The frequency structure of the sidebands is illustrated in Figure 13-4. The center frequency of the carrier wave represented here is 100 MHz. This RF signal is frequency-modulated with a pure tone of 10 kHz. On the high side of the center frequency, there is a sideband for each multiple of the modulating frequency. On the low side of the center frequency, there is a sideband for each submultiple of the modulating frequency. Theoretically these sidebands extend in both directions indefinitely. As the sidebands move out from the center frequency, they gradually grow weaker, but there is no definite pattern as to where they are strong or where they are weak. For best results, all sidebands should be broadcast that are 1 percent of the strength of the center frequency.

### 13.2.3 Bandwidth

The ideal FM bandwidth is wide enough to include all significant sidebands. Even if such a broad band is technically feasible, it is illegal.

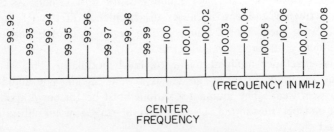

Figure 13-4. FM Sidebands

In the United States, the FM bandwidth is controlled by the Federal Communications Commission. The maximum FM bandwidth is fixed at 150 kHz. Applying this limit to the sideband illustration in Figure 13-4, it is apparent that the legal bandwidth includes seven upper sidebands and seven lower sidebands. All other sidebands must be eliminated regardless of their significance. Any significant sideband that is excluded means a loss of intelligence. This loss shows up in the received signal as distortion.

The percentage of modulation is tied directly to the bandwidth. A deviation of 75 kHz to each side of the center frequency is the legal limit of the 150-kHz bandwidth. By definition, then, a 150-kHz bandwidth is equivalent to 100 percent modulation. The designer's equipment may be capable of much more, but this is all that he is allowed to use.

## 13.3 Frequency Multipliers

In most cases, the RF signal is generated at a frequency much lower than the frequency of transmission. One reason for this is that greater stability in an oscillator can be obtained at the lower frequency. Many designers prefer RF oscillators that are crystal-controlled. Crystal-controlled oscillators have an extremely high stability, but crystals simply are not capable of oscillating in the high-megahertz frequency range. These conditions make it necessary to employ one or more stages of frequency multiplication between the RF oscillator and the final RF power amplifier.

### 13.3.1 Sample Frequency Multiplier

The *frequency multiplier* can be any stage of RF amplification. In fact, the physical layout is no different from any other RF amplifier. Any RF amplifier becomes a frequency multiplier when the output is tuned to a *harmonic* of the input. It is possible to build a frequency quadrupler, but it is not common practice to do so. A quadrupler has too much circuit loss. Better results can be obtained by using two frequency doublers. The basic circuit for two frequency doublers is illustrated in Figure 13-5.

A frequency doubler is easily constructed. The output tank circuit is simply tuned to the second harmonic of the input. This tuning is illustrated in Figure 13-5. The secondary of $T1$ is tuned to the input frequency of 50 MHz. Both windings of $T2$ are tuned to the second harmonic of the 50-MHz input. The secondary of $T2$ feeds 100 MHz to the base of $Q2$. Both windings of $T3$ are tuned to the second harmonic

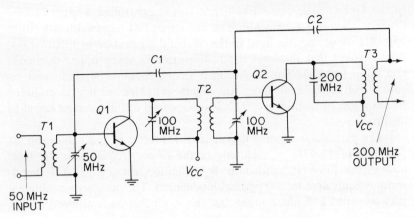

Figure 13-5. Two Frequency Doublers

of $Q2$'s input. A 50-MHz input to $Q1$ produces a 200-MHz output from $Q2$.

The feedback is provided by $C1$ for $Q1$ and $C2$ for $Q2$. The feedback connections at the top of the secondary as indicated here may be too much feedback. Transformers can be obtained with taps on the secondary to enable selection of the proper amplitude for feedback.

### 13.3.2 Efficiency and Power

The efficiency of a frequency multiplier is less than that of a standard amplifier. A standard RF amplifier has an efficiency of 65 to 70 percent. This efficiency drops off to about 42 percent for a second harmonic amplifier (frequency doubler). For a fourth harmonic amplifier (frequency quadrupler) the efficiency is down to about 21 percent.

The power gain indicated on the data sheet may very likely change when applied to a frequency multiplier. Some data sheets show power gain for the fundamental frequency and indicate a power derating factor for the second harmonic. The minimum power-gain factor should be considered in any power calculations for frequency multipliers.

## 13.4 Mixers

The process of mixing two frequencies together and extracting a third frequency is *heterodyning*. In communications receivers, this heterodyning process is used to reduce the received RF carrier wave to a lower, more manageable frequency. A *mixer stage* heterodynes the received RF carrier wave with a locally generated RF signal. This produces a variety of frequencies, including the sum and the difference of the two frequencies. A tuned circuit is used to extract the difference frequency, which still retains all the intelligence of the received RF carrier.

### 13.4.1 Heterodyning

The actual mixing process in heterodyning is very similar to the action of a modulator. The frequency of the received RF carrier wave is determined by the setting of the station selector control. This same control sets the frequency of a local RF oscillator to the incoming RF plus the desired intermediate frequency (IF). The IF for AM receivers operating in the broadcast band is 455 kHz, and the local oscillator is set 455 kHz above the incoming frequency. The mixing action is illustrated in Figure 13-6. The incoming RF carrier contains the center frequency and two sidebands. The signal represented here is a 1000-kHz carrier wave that is modulated by a pure tone of 5 kHz. The mixer receives a center frequency of 1000 kHz and two sidebands. The upper sideband is 1005 kHz, and the lower sideband is 995 kHz. The local oscillator is tuned to 455 kHz above the center frequency; so it is injecting a steady signal of 1455 kHz into the mixer.

The mixing action produces a great variety of frequencies. The 12 outputs listed are only a few of the many. These 12 are composed of the sum and difference of the local oscillator frequency and the three segments of the carrier wave. The output of the mixer stage is a tuned

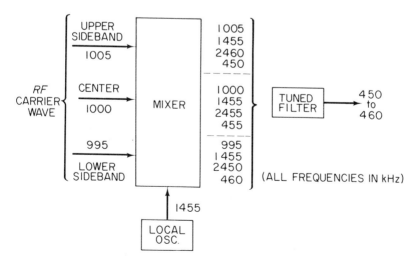

Figure 13-6. Heterodyning

filter circuit that is tuned to the desired intermediate frequency, generally 455 kHz. This filter is a bandpass circuit with a narrow band. It passes frequencies from about 450 to 460 kHz.

### 13.4.2 Base Injection Mixer

Heterodyning can be accomplished by several arrangements. The mixer is generally identified according to the transistor element that receives the local oscillator signal. The base injection mixer has both the received carrier wave and the local oscillator signal applied to the base of the transistor. A sample of such a circuit is illustrated in Figure 13-7. The tuning control varies the capacitance of $C1$ to tune the primary of $T1$ to the selected frequency. If there are preamplifier stages, the same control tunes a similar tank circuit between the antenna and the first preamplifier. The selector control also varies a capacitor in the local oscillator tank circuit to tune the local oscillator to the selector frequency plus 455 kHz.

Transformer $T3$ has a slug-tuned primary which remains set at the center of the IF band, 455 kHz. The transistor responds to all variations on the base, but the instantaneous difference between the two input frequencies is emphasized. The algebraic sum of the amplitude is also a factor in the transistor gain.

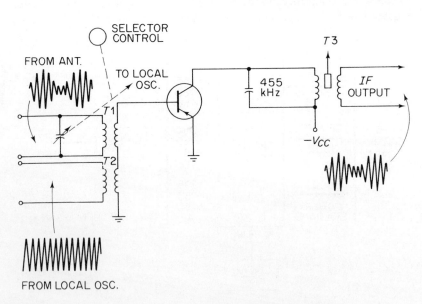

Figure 13-7.   Base Injection Mixer

Modulation and Demodulation

The transistor operates class C, but the tuned circuit in the collector completes each cycle as the transistor cuts off. This tank responds only to 455 kHz and a narrow band which extends above and below this center frequency. The resulting output is an IF signal that is modulated with the same intelligence that was on the carrier wave.

This base injection circuit is sometimes used, but it is not recommended. This circuit arrangement allows the RF carrier wave to feed back from $T1$ to $T2$ and go back into the oscillator. This feedback interferes with the local oscillator frequency. By the same token, the local oscillator frequency can couple from $T2$ to $T1$ and go out toward the antenna. The local oscillator frequency will be broadcast from the antenna as an interference signal. If this circuit is used, it is recommended that at least one stage of RF preamplification be used between the antenna and the mixer. Normal biasing and decoupling circuits have been omitted from this drawing. They, of course, must be added in the same fashion as discussed for other RF amplifiers.

### 13.4.3  Emitter Injection Mixer

The emitter injection mixer is a much more practical circuit than the base injection mixer. The circuit is just as simple, however. This circuit is illustrated in Figure 13-8. The selector control varies capacitors $C1A$ and $C1B$ to tune the primary of $T1$ and $T2$. Capacitors $C2$ and $C4$ are trimmer capacitors, to prevent a gross mismatch of impedance when the tuning capacitors are set to the minimum value. The primary of

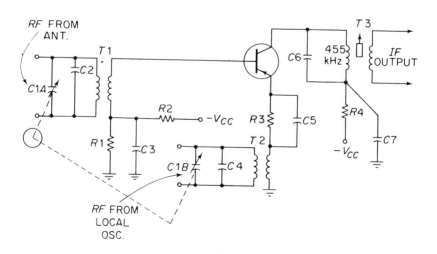

Figure 13-8.  Emitter Injection Mixer

$T1$ is tuned to the center frequency of the desired received carrier-wave frequency, and the primary of $T2$ is tuned 455 kHz above this received RF. The secondary of $T1$ is matched to the input impedance of the transistor. The secondary of $T2$ is an RF load impedance in the emitter of the transistor.

The biasing circuit is formed by $R1$, $R2$, and $R3$. Resistors $R1$ and $R2$ form a voltage divider. Resistor $R3$ is a swamping resistor for stability. Capacitor $C5$ is an RF bypass capacitor to prevent signal degeneration across $R3$.

The tuned tank circuit in the emitter is $C6$ in conjunction with the primary of $T3$. This tank is slug-tuned to the center of the intermediate-frequency band, 455 kHz. Transformer $T3$ is a common IF transformer with a slug-tuned primary. This is an RF stage, and, as such, it must be properly decoupled. Two decoupling capacitors have been included for that purpose: $C3$ and $C7$. Resistor $R4$ aids the decoupling action. Additional decoupling is indicated when RF variations can be detected on the $V_{CC}$ lines.

The received RF is coupled across $T1$ to the base of the transistor. The RF from the local oscillator is coupled across $T2$ to the emitter of the transistor. The two frequencies are mixed, and the difference frequency is extracted by the tuned tank in the collector.

This circuit arrangement isolates the incoming RF from the local oscillator to remove that interference. It also isolates the local oscillator from the antenna and minimizes the undesirable transmission of the local oscillator frequency. This mixer is suitable for use in either AM or FM receivers. Of course, the values of tuned components are *not* the same for both AM and FM because the received frequencies are different.

## 13.5 Frequency Converters

Some designers prefer to incorporate the mixer and local oscillator into a single stage. Such a stage is called a *frequency converter*. When a frequency converter is used, it is generally the first stage in the receiver, taking the RF carrier wave directly from the antenna.

### 13.5.1 Sample Converter Circuit

A sample frequency-converter circuit suitable for use in a low-power AM broadcast band receiver is illustrated in Figure 13-9. Transformer $T2$, with its tuned secondary, is the local oscillator portion of the circuit. The primary of $T2$ is a feedback loop from the transistor to

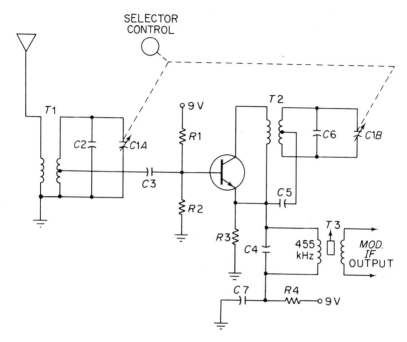

Figure 13-9. Sample Converter Circuit

the tank circuit. This RF feedback is used to sustain oscillation in the tank circuit. The locally produced RF is tapped from the secondary of $T2$ and coupled across $C5$ to the emitter of the transistor.

The RF carrier from the antenna couples across $T1$ to the base of the transistor. The mixing action is as previously described. The 455-kHz tank circuit passes a narrow IF band. This IF has the same modulation as the incoming RF carrier wave. Decoupling is provided by $C7$ and $R4$.

### 13.5.2 Selecting Trial Components

The secondary of $T1$ must be tunable over the entire AM band, and the secondary of $T3$ must be capable of tracking 455 kHz above the setting of $T1$ over the entire range. The taps on these two secondaries should be variable so that the amplitude of the incoming RF can be matched to the amplitude of the oscillations from the $T3$ tank circuit. Resistor $R3$ is the input coupling from $T3$. The value of $R3$ should match the impedance of the transformer tap and provide a small degenerative feedback. A 3000- or 4000-Ω resistor will do for a trial value. The voltage divider, $R1$ and $R2$ in conjunction with $R3$, should set the bias for the desired operating point. If 4 kΩ is used for $R3$, then $R1$ should be 35 kΩ, and $R2$ should be 7 kΩ. The two coupling capacitors, $C3$ and $C5$, are not critical because the frequencies are relatively high; 0.01-$\mu$F capacitors will do nicely. The decoupling capacitor, $C7$, may also be 0.01 $\mu$F. A 0.01-$\mu$F capacitor offers about 35 Ω of resistance to the 455-kHz IF signal. This is adequate decoupling. The decoupling resistor, $R4$, should be about 2 kΩ.

## 13.6 Demodulators

After the signal passes the mixer stage, the IF contains all the intelligence. The same modulation that was on the carrier wave (either AM or FM) has been transferred to the IF signal. After several stages of IF amplification, the intelligence must be extracted from the IF signal. For an AM signal, the process of extracting the intelligence is *rectification* and *filtering*. This rectifying and filtering is performed by a *detector circuit*. For an FM signal the process of extracting the intelligence is to convert the frequency variations to voltage amplitude variations. This conversion action is performed by a frequency *discriminator circuit*.

In either case, the process of extracting intelligence from a modulated wave is *demodulation*. So the AM demodulator is a detector, and the FM demodulator is a discriminator. Either type of demodulator removes the IF signal and leaves an audio envelope. This audio signal should be exactly the same as the modulating signal in the transmitter.

### 13.6.1 Series Diode Detector

The first task in AM demodulation is to clip off half of the IF waveshape through a process of rectification. When the IF component is filtered from the remaining half, only the audio signal remains. The simplest form of AM detector is probably the series diode detector, as illustrated in Figure 13-10.

The basic circuit is Figure 13-10a, while Figures 13-10b, c, and d represent the input, intermediate, and output waveshapes. The primary of $T1$ should be tuned to the 455-kHz center frequency of the IF band. The amplitude-modulated IF signal couples across $T1$, but $CR1$ allows current only on the positive alternations. If capacitor $C2$ is removed from the circuit, the output is similar to waveshape c.

With the filter capacitor, $C2$, in place, rectifying and filtering occur at the same time. Each positive alternation of the IF input draws current from ground through $R1$ and through $CR1$. Current is also drawn from $C2$, charging it with a positive at the top and negative on the

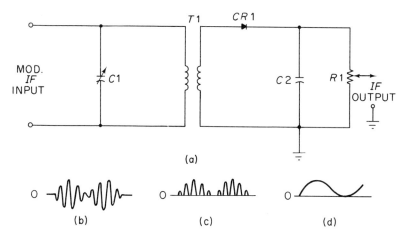

Figure 13-10. Series Diode Detector

ground side. During each negative alternation of the IF, the capacitor is discharging through $R1$. Current from the capacitor is also upward through $R1$, keeping a positive potential at the output. The output audio-frequency signal is varying at the same rate as the modulation on the IF signal.

The amplitude of the AF signal can be varied by the setting of the wiper arm on $R1$. This is a potentiometer, and it is the volume control for the receiver.

The voltage and current requirements for this diode are very modest. The maximum peak voltage from $T1$ is likely less than 2 V. The value of $R1$ should be at least 5 k$\Omega$. So $CR1$ will never have more than 1 mA of current. The filter capacitor, $C2$, should effectively short out the IF signal and still provide a short $RC$ time constant (with $R1$) to the audio frequency. As a trial value, $C2$ should be 0.02 $\mu$F.

The value of $R1$ should match the impedance of $T1$'s secondary and the input impedance of the next stage. Generally it is impossible to accurately match both of these impedances. So the value of $R1$ is a compromise. Since $R1$ is also a volume control, it presents a variable impedance to the next stage. The best results can be obtained when the total resistance of $R1$ matches the secondary impedance of $T1$, and the volume control at midsetting matches the input impedance of the next stage.

### 13.6.2 Shunt Diode Detector

The shunt diode detector is constructed by placing a rectifying diode in parallel with the load resistor. The shunt diode detector normally has a series inductor for filtering. The basic circuit is illustrated in Figure 13-11. The IF signal is rectified by the diode conducting on each negative alternation. Each positive alternation causes current upward through the resistor and through the coil, $L1$. The inductive action of $L1$ keeps current through $R1$ in the same direction during the negative alternations of the IF signal.

Considerations for component values are approximately the same as for the series diode detector. The inductor, $L1$, should offer a high impedance to 455 kHz and a low impedance to the audio signal.

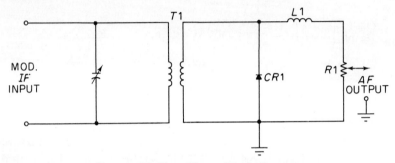

Figure 13-11.  Shunt Diode Detector

### 13.6.3  Transistor Detector

The losses in diode detectors are very high. In fact, diode detectors are seldom more than 50 percent efficient. This low efficiency is of no consequence when detecting a strong signal. If the signal is weak, the losses in a diode detector may be intolerable. The transistor detector can extract the intelligence from a modulated IF and provide signal amplification at the same time. Such a detector is illustrated in Figure 13-12. This is very simply a class C IF amplifier with an $RC$ filter in

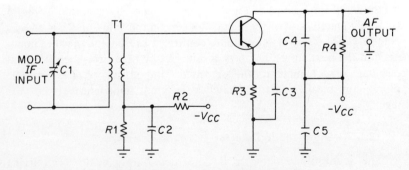

Figure 13-12.  Transistor Detector

# Modulation and Demodulation

the collector circuit. The transistor is biased almost, but not quite, at cutoff. Each positive alternation of the input IF causes the transistor to conduct. Each negative alternation causes the transistor to cut off. When the transistor conducts, current is upward through $R4$. Current is still upward through $R4$ when the transistor is cut off due to the discharge action of $C4$. Bypass capacitor $C5$ shorts all RF components to ground. The output across $R4$ is the audio signal. The output resistor may be a potentiometer volume control.

### 13.6.4 Foster Seeley Discriminator

The discriminator circuit has a complex transformer arrangement, but the remainder of the circuit is very simple. The FM discriminator transformer can be purchased as a complete unit with capacitors. The overall function of the discriminator is to convert the frequency modulation into pulsating dc and change the pulsating dc into an audio-signal equivalent of the frequency modulation. A sample Foster Seeley discriminator is illustrated in Figure 13-13. The enclosed section is $T1$, and it is a complete FM discriminator transformer. The remainder of the circuit resembles two parallel detectors. These twin circuits perform identical functions and should be matched in every way. The characteristics of $CR1$ should be identical to those of $CR2$. The values of $R1$ and $R2$ should be the same, and precision resistors should be used. The capacitance of $C1$ should be the same as that of $C2$, and again precision units are recommended. Remember that any mismatch between the two halves of this circuit will cause distortion.

The voltage and current ratings for each diode are reasonably low.

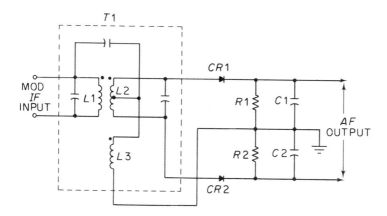

Figure 13-13. Foster Seeley Discriminator

A voltage rating of 2 V peak inverse voltage and a current rating of 1 mA will suffice. The total resistance of $R1 + R2$ should match the impedance of the next stage. Often there is a 450-k$\Omega$ volume control on the input of the next stage. In this case, the sum of $R1$ and $R2$ should be reasonably close to 450 k$\Omega$. Two standard resistors of 200 k$\Omega$ each will be close enough.

The center of the IF band with frequency modulation is 10.7 MHz. Capacitors $C1$ and $C2$ should offer a near short to this IF but should be reasonably large impedances for the audio signal. Capacitors of 550 pF each should be satisfactory when $R1$ and $R2$ are 220 k$\Omega$ each.

The IF transformer, $T1$, is pretuned to the 10.7-MHz IF center frequency. The frequency-modulated IF from the primary of $T1$ is transformer-coupled to the secondary winding ($L2$) and to the tertiary winding ($L3$). The same IF from the primary is capacitively coupled to the center tap of $L2$ and to the top of $L3$. The tertiary winding, $L3$, is also directly coupled to the center tap of $L2$. The bottom of $L3$ couples directly to the center connection of $R1$ and $R2$ as well as $C1$ and $C2$. This point in the circuit is also grounded.

Technically $L3$ is connected across the primary, so the modulated IF is across it at all times. The potential across $L3$ is applied across both diodes, and at the center frequency (10.7 MHz) the diodes conduct equally. When $CR1$ and $CR2$ conduct equally, the current through $R1$ and $R2$ is equal and opposite and produces a zero output.

The diodes have a second voltage from the secondary winding ($L2$) of $T1$. This voltage is 180 degrees out of phase on the two anodes. The diodes are then influenced by the voltage of both $L2$ and $L3$. The level of conduction is determined by the resultant of the two voltages.

At the center frequency, when the IF signal swings positive, $L3$ places a positive on the anodes of both diodes. At the same time, $L2$ places a positive on the anode of $CR1$ and a negative on the anode of $CR2$. The resultant voltage causes $CR1$ to conduct heavily and $CR2$ to conduct very little. This produces current from ground through $R1$ which constitutes a positive output.

Still at the center frequency, when the IF signal swings negative, the situation reverses. Diode $CR2$ now conducts heavily and $CR1$ only slightly. The majority current is now from ground through $R2$, which produces a negative output.

At frequencies above 10.7 MHz, the reactance of $L2$ and $L3$ increases while the associated capacitive reactance decreases. This enables the $L3$ voltage to be applied ahead of the $L2$ voltage and reduces the amplitude of the $L2$ voltage. The action is now determined primarily between the charges on $C1$ and $C2$ and the voltage of $L3$. The level of the difference is determined by the frequency.

## Modulation and Demodulation

At frequencies below 10.7 MHz, inductive reactance has decreased while capacitive reactance has increased. The voltage of $L2$ is now the stronger, and it appears slightly ahead of the voltage of $L3$. Now the $L2$ voltage and capacitor charges determine the action.

The differential voltage across the $R1 + R2$ combination is the audio-frequency output. The filter action of $C1$ and $C2$ removes all the IF component.

### 13.6.5 Ratio Detector

At first glance the ratio detector is very similar to a Foster Seeley discriminator. A brief description should suffice to clarify the differences. Compare the Foster Seeley discriminator in Figure 13-13 with the ratio detector in Figure 13-14. The transformer, $T1$, is a standard FM discriminator transformer with coupling as described for the Foster Seeley discriminator. The external circuit is slightly different. The diodes are oriented in opposite directions; resistors, capacitors, and $L3$ are connected approximately as before, but now the output is taken from the center point of $C1$ and $C2$. Capacitor $C3$ is a filter across the complete $RC$ combination.

The output load resistor for this circuit is $R2$. The resistance of $R2$ should match the impedance of the next circuit. The resistance of $R1$

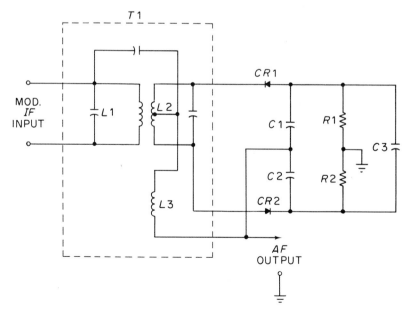

Figure 13-14. Ratio Detector

should be the same as that of $R2$. Capacitors $C1$ and $C2$ should be a near short for 10.7 MHz. Capacitor $C3$ should be of a much higher capacitance than either $C1$ or $C2$. The capacitance of $C1$ and $C2$ can range from 550 pF to 1000 pF, so $C3$ should be on the order of 5 $\mu$F.

The combination $C1$, $C2$, $R1$, $R2$, and $C3$ form a bridge circuit. The frequency of the input at any instant determines the current through the resistors because the frequency controls the ratio of the charges on the capacitors. The equivalent bridge circuit is represented in Figure 13-15. The polarities indicated on the bridge exist at the center fre-

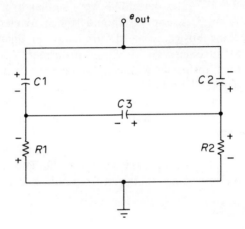

Figure 13-15. Equivalent of a Detector Bridge

quency and hold for frequencies both above and below the center frequency. At the *center* frequency, there is an equal voltage across $C1$, $C2$, $R1$, and $R2$. With this balanced ratio there is zero voltage output, as it should be at the center frequency. As the frequency changes, the ratio changes, and the output changes. At frequencies *above* 10.7 MHz, the output is positive; $C1$ has more charge than $C2$. At frequencies *below* 10.7 MHz, the output is negative; $C2$ has more charge than $C1$. The difference between the charge on $C1$ and the charge on $C2$ is the instantaneous amplitude of the AF signal output.

SECTION FIVE

# SIGNAL-GENERATING CIRCUITS

*Signal-generating circuits* form a big segment of all electronic circuits. Electronic circuits must have the proper operating power and the proper input information. Signal-generating circuits help to fill this need. Electronic circuits require alternating currents at frequencies from a few hertz to several THz (THz = $10^{12}$). The low frequencies are adequately supplied by ordinary electric generators with rotating armatures.

When the frequency needs exceed the practical limits of mechanical rotating devices, electronic generators take the place of mechanical generators. An electronic generator is called an *oscillator*. These oscillators come in two general types: sinusoidal and relaxation. As the name implies, the sinusoidal oscillators produce ac signals that are in the form of sine waves. The relaxation oscillators produce a variety of waveshapes, such as square waves, sawtooth waves, and trigger pulses. The relaxation oscillator may be considered a nonsinusoidal oscillator. The sinusoidal oscillators are discussed in Chapter Fourteen, and Chapter Fifteen is devoted to relaxation oscillators.

CHAPTER FOURTEEN

# Sinusoidal Oscillators

*Sinusoidal oscillators* are used to generate usable sine-wave currents at frequencies that range from low audio to visible light. An oscillator is actually a dc-to-ac converter. It converts direct current to alternating current at a specified frequency. The frequency of oscillation is determined when the designer selects the values of the circuit components.

Sinusoidal oscillators serve many purposes. They are used in modulation circuits, demodulation circuits, and signal generator circuits, to name just a few.

## 14.1 Principles of Oscillation

Any amplifier stage will oscillate if it is given an opportunity to do so. The trick is to sustain oscillation at the desired frequency and, of course, to have the oscillator produce signals of the proper shape and amplitude. The requirements for sustained oscillation are simple: an amplifier stage with a frequency-determining device and regenerative feedback.

### 14.1.1 Amplifying Device

A transistor can be used as the amplifying element for an oscillator at frequencies between the cutoff frequency and the maximum frequency. The cutoff frequency is the frequency that produces a current gain of 0.707 times the low-frequency value, and the low-frequency value is normally measured at 1000 Hz. The maximum frequency ($f_{max}$) is the frequency that reduces the transistor power gain to unity. This explanation of $f_{max}$ applies to the common-emitter configuration. This is not to imply that other configurations *cannot* be used—they can and they are—but the common emitter is by far the most popular.

The common-emitter configuration has moderate input and output impedances that reduce requirements for impedance matching in the feedback circuit. The current and voltage gains are reasonably high, and the power gain exceeds the capabilities of both common base and common collector. There is a phase shift of 180 degrees between input and output that must be planned for in the feedback circuit, but this is a detail that causes no problem.

### 14.1.2 Frequency-Determining Device

The frequency-determining device may be a crystal circuit, an *LC* tank circuit, a bridge, or an *RC* network. The crystal-controlled oscillator operates at the natural frequency of the crystal or at some harmonic of this fundamental frequency. The *LC* tank circuit works on the flywheel effect and has a resonant frequency that is dependent upon the relative values of the inductor and capacitor.

$$f_r = \frac{1}{2\pi\sqrt{LC}}$$

The lag line oscillator uses an *RC* phase-shifting network for the feedback. The operating frequency has 180 degrees of phase shift through the *RC* network. The bridge oscillator generally has an *RC* bridge with frequency determined by the values of *R* and *C*.

### 14.1.3 Feedback

The feedback must be regenerative in nature and strong enough to replace all the energy losses in the circuit. The amplifier power is divided into usable output signal and feedback signal. To avoid excessive losses in the feedback network, its input should match the output impedance of the transistor, and its output should match the input impedance of the transistor.

Phase shift in the network must be planned. If a common emitter is used, the feedback network must shift the phase by 180 degrees to compensate for the 180-degree phase shift through the transistor. If common base or common collector is used, the network must have no phase shift. The same considerations must be given for amplifying devices other than transistors. The signal being fed back, in any case, must be in phase with the input signal.

### 14.1.4 Frequency Stability

A steady fixed frequency of oscillation is a prime requisite for most oscillators. Many factors can affect the stability of this frequency. The dc operating point must be fixed on a linear portion of the amplifier's characteristics curve. Nonlinear operation will cause parameter changes and frequency shifts. Not only must the operating point be set on the linear part of the curve, it must not shift for any reason. These requirements make it essential to have a regulated voltage source for the dc bias on the amplifier. Temperature stabilization is also important, since changes in temperature tend to cause a shift of the operating point.

Internal capacitance must be considered. In a common-emitter circuit, the collector-to-emitter capacitance can be bothersome. This capacitance changes with frequency, voltage, or temperature changes. The capacitance is more troublesome as the frequency increases. At high frequencies, it may be necessary to place an external capacitor between collector and emitter. This external capacitance is in parallel with the internal capacitance, and this increases the capacitance and reduces the capacitive reactance.

## 14.2 Keyed Audio Oscillator

A *keyed audio oscillator* is a very simple circuit, and yet it incorporates all the principles of the more complex oscillators. The objective is to build a simple portable circuit that will produce an audible tone in a headphone when a key is pressed.

### 14.2.1 Sample Circuit

Consider the headphone to be a magnetic type which contains an inductor. Using the inductor in the headphone as a start, capacitors can be added to form a resonant circuit. A dry cell should be used for portability, a transistor for amplification, a keying device for turning the tone on and off, and a variable resistor for volume control. Putting these items together and adding a few necessary resistors produces the circuit in Figure 14-1.

The biasing circuit is composed of $R1$, $R2$, and $R3$. Capacitors $C1$ and $C2$ form a resonant circuit with the inductor in the headphone. Resistor $R3$ provides stability and produces a small degenerative feedback. Regenerative feedback from the resonant circuit is applied directly to the emitter with no phase shift. Resistor R4 controls the collector current and can be set for the desired volume in the headphone.

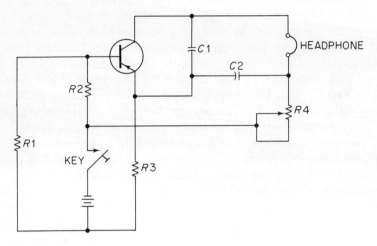

Figure 14-1.  Keyed Audio Oscillator

### 14.2.2  Selecting Component Values

The circuit must be designed around the headphone. Assuming a 2000-Ω magnetic headphone, the inductance is approximately 320 mH. This 2000 Ω is true only for a frequency of 1000 Hz. A good tone can be produced at a frequency of about 1200 Hz, and at this frequency $X_L$ is 2400 Ω. The value of C1 and C2 should be selected to form a tank that is resonant to this 1200-Hz frequency. Since $X_L$ is 2400 Ω, $X_C$ has to be 2400 Ω, so total capacitance of C1 in series with C2 is calculated by

$$C = \frac{1}{2\pi f X_C}$$

At 1200 Hz this is a total capacitance of about 0.06 μF. Putting two 0.1-μF capacitors in series produces a total capacitance of 0.05 μF. This is close enough; so C1 and C2 should be 0.1 μF each.

The battery may be any dry cell with a voltage from 1.5 V to 4.5 V. The higher the voltage, the higher the possible volume.

As trial values let $R1 = 2200$ Ω, $R2 = 2700$ Ω, and $R3 = 3000$ Ω. The volume control should have a wide range, but its value is not critical. Resistor R4 should be a potentiometer with a total resistance of at least 5000 Ω. All resistors should have a wattage rating of 0.5 W.

## 14.3 Twin-T Audio Oscillator

The *twin-T oscillator* uses an *RC* network for determining the frequency. The name is taken from the arrangement of capacitors and resistors in the *RC* network. The twin-T oscillator can be constructed to oscillate at any given frequency from 2 Hz to 2000 kHz. Normally it is built to produce a signal in the audio-frequency band. The output from a twin-T oscillator is a highly stable frequency that can be used in a number of testing situations. It is especially useful in testing high-fidelity audio equipment and amateur transmitters.

### 14.3.1 Sample Circuit

Only one transistor is needed to construct a twin-T oscillator, but it is advisable to add an isolation stage. Otherwise the output is taken across the frequency network, and the loading effect tends to cause instability. Many arrangements are possible. One functional circuit is illustrated in Figure 14-2. Transistor *Q*1 with its associated circuit is

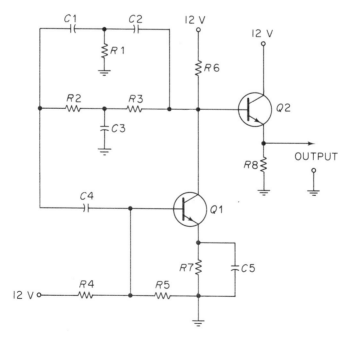

Figure 14-2. Twin-T Audio Oscillator

the oscillator, and $Q2$ is a common-collector isolation and power-amplifier stage.

The twin-T, $RC$-frequency network is composed of $C1$, $C2$, $C3$, $R1$, $R2$, and $R3$. Without this network, $Q1$ is a standard common-emitter amplifier. The bias is provided by $R4$, $R5$, and $R7$. The resistor, $R7$, is bypassed to prevent degeneration. The collector load resistor is $R6$. Capacitor $C4$ is a standard coupling capacitor.

The twin-T network provides 180 degrees of phase shift for the operating frequency, and this constitutes regenerative feedback from collector to base. The oscillator output is directly coupled to the base of $Q2$. The output is developed across the emitter load resistor of $Q2$.

### 14.3.2 Selecting Component Values

Both transistors must be capable of handling the frequency of oscillation, but this is no problem as long as the frequency is in the audio band. At frequencies about 20 kHz, the $f_{max}$ of the transistors may be a factor. Both transistors should operate class A and therefore require forward bias. The following resistor values will serve as a starting point:

$$R4 = 100 \text{ k}\Omega$$
$$R5 = 6800 \text{ }\Omega$$
$$R7 = 2200 \text{ }\Omega$$

A twin-T oscillator has a lot of losses, so both transistors should produce maximum gain to offset this deficiency. The value of the collector load resistor of $Q1$, $R6$, should be about 60 times as great as the resistance of $R7$. With $R7$ set at 2200 $\Omega$, the value of $R6$ is indicated as 132 k$\Omega$. The closest standard resistor is satisfactory, so $R6$ becomes 120 k$\Omega$. The output load resistor, $R8$, must depend to some extent on what the output is coupled to. If it is specified that this circuit feed loads *not less* than 3000 $\Omega$, the value of $R8$ can be set to 820 $\Omega$.

In the twin-T circuit, the resistance of $R1$ should be slightly larger than that of $R7$. With $R7$ of 2200 $\Omega$, $R1$ should be about 2700 $\Omega$. The resistance of $R2$ should be 10 times the resistance of $R1$, and $R2$ and $R3$ should be the same value. So $R2$ and $R3$ each become 27 k$\Omega$.

The value of all capacitors are frequency-dependent, so what is the desired operating frequency? When this has been decided, the following logic can be used to determine the values of $C1$, $C2$, and $C3$.

1. $C1$ must be equal to $C2$.
2. $C3$ is equal to $C1 + C2$.
3. $f = 1/5RC$ and $C = 1/5Rf$ (where $R = R1$ or $R2$ and $C = C1$ or $C2$).

Sinusoidal Oscillators

These equations normally contain $2\pi$ in the denominator instead of the 5 shown here. This slightly smaller denominator creates an out-of-balance condition that enhances oscillation. Using this logic, if the desired operating frequency is 10 kHz, $C1$ and $C2$ are each 500 pF; for 6 kHz they are each 1000 pF; for 600 Hz they are 0.01 $\mu$F; and for 100 Hz they become 0.05 $\mu$F. In each case, $C3$ is the sum of $C1$ plus $C2$.

Capacitor $C4$ is a coupling capacitor and should be a low impedance in respect to the value of $R5$. Since $R5$ is 6800 $\Omega$, a 1-$\mu$F capacitor should be sufficient coupling for any desired frequency. Capacitor $C5$ should be a virtual short with respect to the 2200 $\Omega$ of $R7$. For operating frequencies up to 2 kHz, $C5$ should be 300 $\mu$F; for frequencies above 2 kHz, $C5$ should be 5 $\mu$F. All capacitors should be rated 6 V dc except $C4$; $C4$ should be rated at 12 V.

All resistors should have a power rating of at least 0.5 W. It may be helpful in obtaining a stable oscillating condition with a slight adjustment in the feedback circuit. Resistor $R1$ can be a variable resistor for this purpose. If stable oscillation is difficult to achieve, the gain of $Q1$ must be increased. Increased gain can be obtained by selecting a transistor with a higher beta or by increasing the resistance of $R6$.

## 14.4 Lag-Line Oscillator

*Lag-line oscillators* are good audio oscillators, and they avoid the use of inductors. The resonant $LC$ tank circuit is *not* a practical circuit in audio oscillators. The lag-line oscillator works on the principle of an $RC$ phase-shifting feedback network. The feedback circuit is also the frequency-determining device.

### 14.4.1 Sample Circuit

The lag-line oscillator is an ordinary amplifier with a ladder-like feedback circuit. A sample circuit is illustrated in Figure 14-3. Bias for the transistor is provided by $R3$, $R4$, and $R5$. Resistors $R3$ and $R6$ serve a dual purpose. One leg of the feedback circuit is composed of $C3$ and $R3$. Transistor swamping is provided by $R5$. The bypass capacitor, $C4$, prevents signal degeneration. The collector load resistor is $R6$.

Feedback and frequency determination takes place in the three-section $RC$ network between collector and base. Each $RC$ segment is capable of shifting the phase of a signal by a maximum of 90 degrees. Better results are obtained by using three segments and obtaining a 60-degree phase shift in each segment.

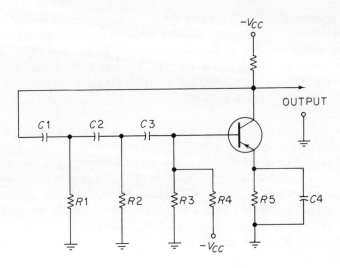

Figure 14-3. Lag-Line Oscillator

Oscillation is started by applying power to the circuit. At the first instant, there is a surge of noise through the transistor. This noise contains frequencies that range across the entire spectrum. One of these frequencies will feed back through the RC network and have its phase shifted by exactly 180 degrees. This frequency is quickly reinforced until it becomes the dominant frequency. This dominant frequency is the frequency of operation.

### 14.4.2 Selecting Component Values

Energy losses in a lag-line oscillator are relatively high. The high loss makes a high-gain transistor essential to proper operation. The transistor must have a gain of at least 60 at the operating point. The collector swing of the transistor determines the amplitude of the output voltage, and the values of $V_{CC}$ and $R6$ are selected according to the desired output. This collector load resistor should be about equal in value to either $R1$, $R2$, or $R3$ in order to allow the maximum collector swing. The bias should be set for a $V_c$ equal to 0.5 $V_{CC}$ without feedback.

The feedback network is composed of three equal RC segments: $C1R1$, $C2R2$, and $C3R3$. All three resistors should have equal resistance, and all three capacitors should have equal capacitance. The relative values of capacitance and resistance for each segment should be such that the operating frequency is shifted by 60 degrees. This 60-degree shift is realized when $X_C = 1.6\,R$.

The value of $R1$, $R2$, and $R3$ should be set to 10 kΩ each as a trial value. The desired operating frequency will then determine the selected value for each of the capacitors. For a frequency of 1000 Hz, $R = 10$ k$\Omega$ and $X_C = 16$ k$\Omega$. At 1000 Hz a 0.01-$\mu$F capacitor will have an impedance very near 16 k$\Omega$. So these values should be sufficient, if the desired frequency of oscillation is indeed 1000 Hz.

$$R1 = R2 = R3 = 10 \text{ k}\Omega$$
$$C1 = C2 = C3 = 0.01 \text{ } \mu\text{F}$$

These resistors and capacitors should be precision components, but even with precision components, there is still tolerance. It is wise to have a variable component in one or more of the $RC$ segments to enable adjustment for drift and tolerance of components. Some designers prefer to have adjustable components in all three segments. The same results can be obtained by varying either resistance or capacitance. If a variable frequency is desirable, it is recommended that a three-section variable capacitor be used.

## 14.5 Bipolar Colpitts RF Oscillator

*Colpitts oscillators* are widely used as RF oscillators in transmitters, receiver local oscillators, and test equipment. They can be designed to operate at any given frequency or to be tunable over a wide band of frequencies. A well-designed Colpitts oscillator can produce highly stable frequencies up to 144 MHz.

### 14.5.1 Sample Circuit

The Colpitts oscillator may use either a bipolar transistor or a FET. The bipolar design utilizes both common base and common collector, and the tank circuit may be located in either the base or the collector circuit. The sample circuit in Figure 14-4 is a common emitter with a tuned collector. The identification mark of the Colpitts oscillator is the double capacitor ($C1$ and $C4$) across the inductor in the tank circuit and the feedback connection between the two capacitors. This tank circuit determines the frequency of this oscillator.

The biasing circuit is formed by $R1$, $R2$, and $R4$. This should be class A bias, and it should be designed so that the reverse bias across $R4$ is about 50 percent of the feedback signal.

Capacitor $C1$ is a decoupling capacitor. It should offer a virtual short to the lowest operating frequency. Other decoupling devices are obviously required, but they have been omitted for the sake of clarity.

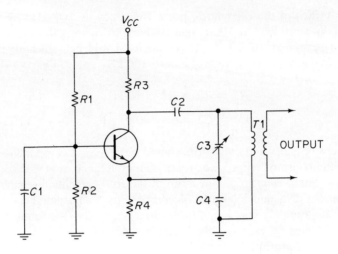

Figure 14-4. Bipolar Colpitts RF Oscillator

The resonant tank circuit is formed by $C3$, $C4$, and the primary winding of $T1$. The resonant frequency of this tank circuit is the frequency of oscillation. The variable capacitor, $C3$, indicates that the frequency can be varied. Capacitor $C2$, in conjunction with $C3$ and $C4$, forms a voltage divider to compensate for the interelement capacitance of the transistor.

The output is coupled to the secondary of $T1$. This is a proper output for a low impedance load. For high impedance loads, a capacitor output may be connected directly to the top of the $T1$ primary.

### 14.5.2 Selecting Component Values

As trial values for the resistors, $R3$ should be such that $V_C = 0.5\ V_{CC}$ at the operating point. In other words, the static collector current drops 50 percent of the supply voltage across $R3$. The value of $R2$ should be approximately the same as that of $R3$, and $R4$ should be no more than 50 percent of $R3$. With a high-gain transistor, the starting resistor values should be:

$$R1 = 6.4\ \text{k}\Omega$$
$$R2 = 2\ \text{k}\Omega$$
$$R3 = 2\ \text{k}\Omega$$
$$R4 = 1\ \text{k}\Omega$$

# Sinusoidal Oscillators

These values will likely need some adjustment when the circuit is bench tested.

Capacitor $C2$ is a series dropping impedance for signals into the tank circuit. This capacitor should offer no more than 200 Ω of capacitive reactance to the lowest frequency of oscillation. For example: if the low frequency is 100 kHz, then

$$C2 = \frac{1}{2\pi f X_C}$$

$$= \frac{0.159}{100 \times 10^3 \times 200}$$

$$= \frac{0.0159 \times 10^{-6}}{2}$$

$$= 0.008 \; \mu F$$

This is the minimum capacitance, and a slightly larger capacitance is satisfactory.

The resonant tank circuit is the critical point in the circuit. This circuit is formed of $C3$, $C4$, and the primary of $T1$. The best procedure is probably to first select a transformer then design around it. Suppose that the primary winding of $T1$ has an inductance of 2 mH. The total capacitance can then be calculated using the lowest frequency as the resonant frequency.

The inductive reactance and capacitive reactance are the same: $X_C = X_L$ and

$$X_L = 2\pi f L$$

At 100 kHz minimum frequency this equates to 1256 Ω for both $X_L$ and $X_C$. The total capacitance then is

$$C = \frac{1}{2\pi f X_C}$$

For a frequency of 100 kHz that means that the $C3C4$ combination is approximately 0.001 $\mu F$. This is the maximum capacitance setting for $C3$. As the capacitance of $C3$ is decreased, the resonant frequency of the tank circuit increases.

The total capacitance for $C3$ and $C4$ in series is

$$C_t = \frac{C3 C4}{C3 + C4}$$

The value of $C4$ should be about three times the maximum value of $C3$.

## 14.6 Crystal-Controlled Colpitts RF Oscillator

Most types of RF oscillators can use a crystal as the frequency-controlling device. This is a popular arrangement because of the extremely high stability obtained from the crystal control. The *crystal-controlled oscillator* may be designed to use one crystal and produce one stable frequency of operation. This stable frequency may be the crystal fundamental or some harmonic of the fundamental frequency. This oscillator can also be designed for switching among several stable frequencies by switching in different crystals and matching tank circuits.

### 14.6.1 Sample Circuit

A Colpitts oscillator lends itself readily to crystal control, and, when reduced efficiency can be tolerated, it may be designed to operate at multiples of the crystal frequency up to the fourth harmonic. A well-designed crystal-controlled Colpitts oscillator is about 70 percent efficient when operated at the crystal fundamental frequency. At the second harmonic the efficiency is about 42 percent. On the fourth harmonic the efficiency is only about 21 percent. Figure 14-5 illustrates one arrangement for a crystal-controlled Colpitts RF oscillator.

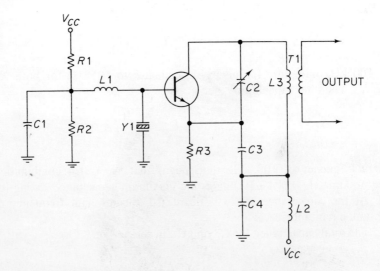

Figure 14-5. Crystal-Controlled Colpitts RF Oscillator

## Sinusoidal Oscillators

This circuit is an amplifier with a tuned base and a tuned collector. The crystal, $Y1$, forms a tuned tank in the base circuit, and this crystal tank controls the frequency. The tuned collector tank is composed of $C2$, $C3$, and the primary winding of $T1$. The resonant frequency of the collector tank determines the operating and output frequency, but this collector tank must be designed to operate on the crystal fundamental frequency or one of its harmonics.

Decoupling of the RF signal is provided by $C1$, $L1$, $C4$, and $L2$. Inductors $L1$ and $L2$ are RF choke coils, while capacitors $C1$ and $C4$ are RF bypass capacitors.

All three resistors are used in biasing the transistor, and $R3$ provides some degeneration while coupling the feedback from the tank to the emitter.

### 14.6.2 Selecting Component Values

The transistor must have a high gain to compensate for the low efficiency of the circuit. This is especially true if the oscillating frequency is some harmonic of the crystal fundamental frequency. The remainder of this discussion assumes the following conditions:

$$\text{Crystal frequency} = 100 \text{ kHz}$$
$$\text{Output frequency} = 400 \text{ kHz}$$
$$\text{Inductance of } L3 = 20 \text{ }\mu\text{H}$$

The inductive reactance of $L3$ is equal to the capacitive reactance at the 400-kHz resonant frequency, so

$$X_L \text{ or } X_C = 2\pi f L$$
$$= 2\pi \times 400 \times 10^3 \times 20 \times 10^{-6}$$
$$= 50 \text{ }\Omega$$

The total capacitance of $C2$ and $C3$ must exhibit 50 $\Omega$ of $X_C$ to the resonant frequency, so

$$C = \frac{1}{2\pi f X_C}$$
$$= \frac{0.159}{400 \times 10^3 \times 50}$$
$$= 0.0079 \text{ }\mu\text{F}$$

A slightly larger capacitance is acceptable, so 0.0079 µF can be rounded off to 0.008 µF. Capacitor C3 should be about three times the capacitance of C2 when C2 is at its maximum capacitance setting.

Capacitor C2 is a variable capacitor, but it is not a frequency control. This oscillator is intended to produce a steady, fixed frequency. The internal collector to emitter capacitance is in parallel with C2. This internal capacitance adds to the value of C2. The adjustment of C2 decreases its capacitance and increases the effect of the internal capacitance.

The decoupling components must decouple the fundamental crystal frequency of 100 kHz. The $X_L$ of L1 and L2 must be fairly high for this frequency, while the $X_C$ of C1 and C4 must be very low. An $X_L$ of 2000 Ω is about right for a choke coil. The inductance of each choke coil should be

$$L = \frac{X_L}{2\pi f}$$

$$= \frac{2000}{6.28 \times 100 \times 10^3}$$

$$= 3 \text{ mH}$$

The decoupling capacitors must be less than 200 Ω at the crystal frequency; 50 Ω is much better.

$$C = \frac{1}{2\pi f X_C}$$

$$= \frac{0.159}{100 \times 10^3 \times 50}$$

$$= 0.03 \text{ µF}$$

As trial values R3 should be 700 Ω and R2 about three times the value of R3. The value of R1 is then about 3700 Ω.

## 14.7 MOSFET Colpitts RF Oscillator

Field-effect transistors are gradually replacing bipolar transistors in most common oscillator circuits. In some respects the *MOSFET Colpitts RF oscillator* makes a superior high-frequency oscillator. One reason for this is the relatively low input capacitance. This input capacitance is on the order of 5 or 6 pF at a frequency of 1 MHz.

### 14.7.1 Sample Circuit

A sample circuit of the MOSFET Colpitts RF type is illustrated in Figure 14-6. This is a variable-frequency RF oscillator with a minimum frequency of 8 MHz. With slight variations of component values, this oscillator is suitable for single-sideband, 50-MHz, and 144-MHz transmitters. The feedback tap between $C4$ and $C5$ identifies this as a Colpitts oscillator. The input capacitance of the MOSFET is compensated for by the voltage-divider arrangement of $C3$, $C4$, and $C5$.

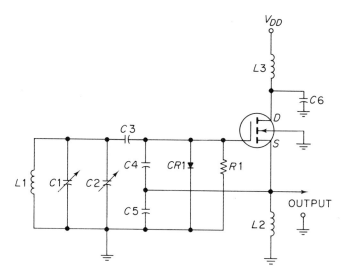

Figure 14-6. MOSFET Colpitts RF Oscillator

The frequency of oscillation is determined by $L1$ and the total capacitance paralleling this coil. Inductors $L2$ and $L3$ are RF choke coils and $C6$ is a decoupling capacitor.

Resistor $R1$ and rectifier $CR1$ provide rectified current for the oscillator. This arrangement provides a degree of automatic bias, which adds substantially to the stability of the frequency.

This oscillator provides a low-level output and should be followed by amplification and isolation stages. Without proper isolation the operating frequency is likely to be unstable.

### 14.7.2 Selecting Component Values

Since the operating frequency is to be 8 MHz and up, the first consideration is the frequency capability of the transistor.

Next the decoupling circuits should be designed. The decoupling must eliminate the lowest frequency from the dc lines. This means that $L2$ and $L3$ should be about 2000 Ω and $C6$ should be about 50 Ω to the low frequency of 8 MHz. The inductance that offers 2000 Ω of $X_L$ to 8 MHz is

$$L = \frac{X_L}{2\pi f}$$

$$= \frac{2000}{6.28 \times 8 \times 10^6}$$

$$= 40 \ \mu H$$

The capacitance of $C6$ that offers an $X_C$ of 50 Ω to 8 MHz is

$$C = \frac{1}{2\pi f X_C}$$

$$= \frac{0.159}{8 \times 10^6 \times 50}$$

$$= 397 \ pF$$

In this case, $C4$ and $C5$ should be equal at about 560 pF each. Capacitor $C3$ should be about 270 pF. Now if the maximum capacitance of $C1$ is set to 50 pF and the maximum capacitance of $C2$ is set to 25 pF, the total capacitance is about 150 pF. The $X_L$ of $L1$ should equal the total $X_C$ at a frequency of 8 mHz, so

$$X_L \text{ or } X_C = \frac{1}{2\pi f C}$$

$$= \frac{0.159}{8 \times 10^6 \times 150 \times 10^{-12}}$$

$$= 132 \ \Omega$$

The inductance of $L1$ should be such that it offers an $X_L$ of 132 Ω to a frequency of 8 MHz.

$$L1 = \frac{X_L}{2\pi f}$$

$$= \frac{132}{6.28 \times 8 \times 10^6}$$

$$= 2.6 \ \mu H$$

This inductor of 2.6 µH will form a resonant circuit with the total capacitance of the circuit. Capacitor $C1$ is the tuning control and $C2$ is an air-dielectric-type trimmer.

## 14.8 Armstrong RF Oscillator

Several versions of the *Armstrong RF oscillator* are popular in transmitter and receiver circuits as well as in test instruments. The identifying mark of an Armstrong oscillator is the tickler coil pickup loop that is used for feedback coupling. In fact, the Armstrong oscillator is called a *tickler-coil oscillator*.

### 14.8.1 Sample Circuit

The tuned collector Armstrong oscillator is an easy circuit to build and is very practical. A sample of such a circuit is illustrated in Figure 14-7. The biasing circuit is formed with $R1$, $R2$, and $R3$. Capacitor $C2$ is an RF bypass to prevent degeneration of the signal across the emitter resistor, $R3$. Decoupling is accomplished by $C1$, $C4$, and RF choke $L1$. The oscillating frequency is determined by the tank circuit, which is composed of $C3$ and $L2$. Inductor $L3$ is a third winding on $T1$; this is the tickler coil that picks up a portion of the RF signal from the tank circuit and feeds it back to the base. The arrangement of $L2$ and $L3$

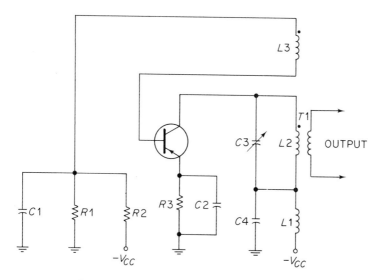

Figure 14-7. Armstrong RF Oscillator

must be such that the feedback to the base is 180 degrees out of phase with the signal on the collector. The frequency of oscillation can be adjusted by varying $C3$.

### 14.8.2 Selecting Component Values

First select a relatively high gain transistor capable of handling the current when the full $V_{CC}$ is on the collector. Make sure that the transistor can handle the frequency range that is expected to be used. The transistor should have a minimum of capacitance between collector and emitter. This output capacitance is in parallel with, and additive to, the capacitance of $C3$. At the minimum frequency, $C3$ is set to its maximum capacitance and the interelectrode capacitance has no appreciable effect. As $C3$ is tuned to increase the frequency, its capacitance decreases, and the effect of interelectrode capacitance increases. When collector-to-emitter capacitance becomes one-tenth of the value of $C3$, these two capacitances must be added in order to determine the resonant frequency of the tank circuit.

In this example it is assumed that the low-frequency limit is 500 kHz and the inductance of $L2$ is 3 mH. At resonance the $X_L$ of $L2$ is equal to the $X_C$ of $C3$ with $C3$ set to its maximum capacitance. Therefore,

$$X_L \text{ or } X_C = 2\pi f L$$
$$= 6.28 \times 500 \times 10^3 \times 3 \times 10^{-3}$$
$$= 9420 \, \Omega$$

Capacitor $C3$ at its maximum capacitance setting must offer an $X_C$ of 9420 $\Omega$ to the minimum frequency of 500 kHz. Therefore,

$$C = \frac{1}{2\pi f X_C}$$
$$= \frac{0.159}{500 \times 10^3 \times 9420}$$
$$= 34 \text{ pF}$$

The decoupling capacitors should offer no more than 50 $\Omega$ of reactance to 500 kHz. So $C1$, $C2$, and $C4$ should each be

$$C = \frac{1}{2\pi f X_C}$$

# Sinusoidal Oscillators

$$= \frac{0.159}{500 \times 10^3 \times 50}$$

$$= 0.006 \ \mu F$$

Inductor $L1$ should offer at least 2000 Ω to 500 kHz. So the inductance of $L1$ is

$$L = \frac{X_L}{2\pi f}$$

$$= \frac{2000}{6.28 \times 500 \times 10^3}$$

$$= .64 \text{ mH}$$

Supposing that $C3$ can be adjusted from 34 pF to 20 pF, the top frequency becomes

$$f = \frac{1}{2\pi\sqrt{LC}}$$

$$= \frac{0.159}{\sqrt{3 \times 10^{-3} \times 20 \times 10^{-12}}}$$

$$= 649 \text{ kHz}$$

## 14.9  Crystal-Controlled Armstrong RF Oscillator

Crystals are sometimes used to stabilize an Armstrong RF oscillator. A slight modification on the circuit of Figure 14-7 will illustrate one method of installing a crystal. The oscillating frequency becomes fixed unless several crystals are used, and even so, the frequency with any given crystal is limited to one fixed frequency.

### 14.9.1  Sample Circuit

The sample circuit in Figure 14-8 is a slight modification of the Armstrong RF oscillator in Figure 14-7. This oscillator has a tuned tank, $Y1$, in the base circuit as well as a tuned tank in the collector circuit. The frequency of $Y1$ is the control frequency, and the collector tank must be tuned to the fundamental or a harmonic of $Y1$.

Capacitor $C3$ is variable, but it is only a fine control. It changes frequency only slightly.

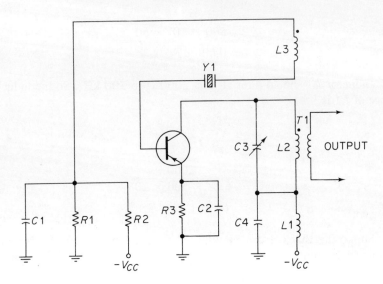

Figure 14-8. Crystal-Controlled Armstrong RF Oscillator

The other components in the circuit perform the same function as described for Figure 14-7.

### 14.9.2 Selecting Component Values

Assuming a fundamental crystal frequency of 500 kHz and an operating frequency the same as the crystal fundamental, all the calculations made for Figure 14-7 are valid for Figure 14-8. The one exception, of course, is that the frequency is *not* variable in this case—but that has already been emphasized.

CHAPTER FIFTEEN

# Relaxation Oscillators

*Relaxation oscillators* are the signal generators for square waves, sawtooth waves, trigger pulses, and time delay gates. In the process of producing these various waveshapes, they become signal control circuits as well as signal generators. Among the many relaxation oscillators, the most common are probably sweep generators, blocking oscillators, and multivibrators.

## 15.1 RC Transients

Most relaxation oscillators depend upon the principle of *transients* in an $RC$ network: the charge and discharge of a capacitor in series with a resistor. The rate of charge or discharge is determined by the relative values of $R$ and $C$.

### 15.1.1 Charge and Discharge Time

The principles of $RC$ transients are illustrated in Figure 15-1. In the present condition the switch is open, the capacitor is fully discharged, and there is no path for current. Moving the switch to position A places the battery in series with the capacitor and resistor and starts current in the circuit.

At the first instant, all the voltage appears across the resistor and maximum current exists. As the capacitor charges, the charge opposes the battery voltage and decreases current. After a time equal to $5RC$, the capacitor charge is equal and opposite to the battery voltage, and the current ceases. The capacitor is now fully charged.

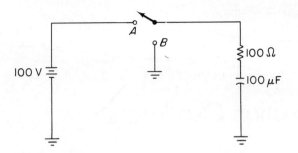

Figure 15-1. Capacitor Charge and Discharge

Moving the switch from A to B removes the battery from the circuit and completes a discharge path for the capacitor. Current exists from the grounded plate of the capacitor, through the switch, and through the resistor to the top plate of the capacitor. At the first instant, the current is maximum with the total charge across the resistor. In this case, the charge is equal to the battery voltage, 100 V. As the capacitor discharges, the current decreases. After a time equal to $5RC$ the capacitor is fully discharged and current ceases.

The components used in this circuit provide a current that ranges from 0 to 1 A and back to zero. The capacitor charges from 0 to 100 V in

$$t = 5RC$$
$$= 5 \times 100 \times 100 \times 10^{-6}$$
$$= 0.05 \text{ s or } 50 \text{ ms}$$

At the end of 50 ms, the full 100 V is on the capacitor. When the switch is changed, the discharge of the capacitor requires exactly the same time as the charge: 50 ms. The quantity of the applied voltage has no bearing on the time for either charge or discharge. The only factor on time is the relative values of $R$ and $C$.

### 15.1.2 Time Constants

An $RC$ time constant is a time equal to $R$ times $C$:

$$tc = RC$$

where $tc$ is in seconds, $R$ is in ohms, and $C$ is in farads. Since the capac-

# Relaxation Oscillators

itor charges or discharges in $5RC$, that is the same as five time constants.

One time constant is also the time required for a capacitor to charge to 63 percent of the applied voltage on the charge cycle or to lose 63 percent of its charge on the discharge cycle. This voltage change per time constant is illustrated in Figure 15-2.

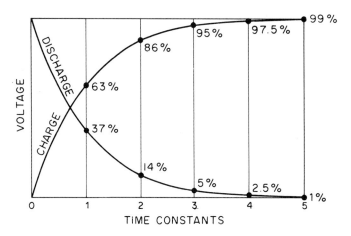

Figure 15-2. Capacitor Voltage Versus Time Constants

## 15.2 Unijunction Sawtooth Generator

The *unijunction sawtooth generator* is identical to the circuit in Figure 15-1 except for a unipolar transistor that replaces the mechanical switch. This is probably the simplest form of relaxation oscillator.

### 15.2.1 Sample Circuit

A sample unijunction sawtooth generator, together with its approximate output waveshape, is illustrated in Figure 15-3. The full battery voltage, $V$, is applied between base 1 and base 2. The silicon bar between the two bases acts as a voltage divider. The voltage gradient in the bar keeps the emitter potential someplace between the potential of $B1$ and $B2$.

When power is applied, there is reverse bias between emitter and both bases. A small reverse current from emitter to base 2 starts charging the capacitor with negative on the grounded plate. As this charge builds up, the emitter becomes more positive. After a short time, the emitter becomes positive with respect to $B1$. Forward bias now exists

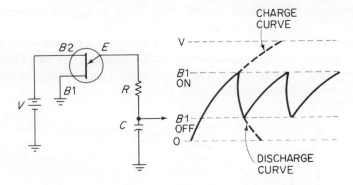

Figure 15-3. Unijunction Sawtooth Generator

between emitter and $B1$. Current from $B1$ to emitter provides a discharge path for the capacitor.

As the capacitor slowly discharges, the voltage across the capacitor and the voltage on the emitter decrease. After a short time, a reverse bias again exists between emitter and $B1$. The forward current from $B1$ to emitter ceases, and reverse current from emitter to $B2$ starts to charge the capacitor again. This is the beginning of the second cycle.

Notice on the waveshape that after the power is first applied, the charge on the capacitor never reaches the applied voltage, $V$, and it never reaches 0 V. The output is taken across the capacitor. The maximum amplitude of the output is limited by turn-on of $B1$. The minimum amplitude of the output is limited by $B1$ turn-off.

The slope of the leading edge of the output waveshape is the capacitor charge curve. The capacitor discharge curve is the slope of the trailing edge. The rise time and fall time are both determined by the relative values of resistance and capacitance. The slope, and consequently the cycle duration, can be made adjustable by using either a variable capacitor or a variable resistor.

### 15.2.2 Selecting Component Values

The first step is to choose a unijunction transistor that has a $B1$ on-potential at the desired maximum output level and $B1$ off-potential at the desired minimum output level. In choosing the transistor, make sure that the available voltage supply exceeds the $B1$ on-potential by a large ratio; a ratio of 10:1 is sometimes needed.

The values of $R$ and $C$ determine rise time, fall time, and cycle time. The value of a time constant is $R \times C$, so changing the value of either $R$ or $C$ changes the duration of the time constant. A great many combinations of resistance and capacitance will produce the desired output. It is recommended that decade devices be used for both $R$ and $C$ in the test circuit. These values can then be adjusted experimentally until the exact desired waveshape is produced across the capacitor. The decade devices can then be replaced with the proper values of $R$ and $C$.

## 15.3 Zener Diode Sawtooth Generator

The *Zener diode sawtooth generator* is another very simple relaxation oscillator with a variety of uses in timing circuits.

### 15.3.1 Sample Circuit

The circuit and output waveshape of a Zener diode sawtooth generator are illustrated in Figure 15-4. When power is applied, current through the resistor starts to charge the capacitor with negative on the grounded plate. As the capacitor charges, point A gradually becomes more positive. A positive potential at point A constitutes reverse bias for the Zener diode.

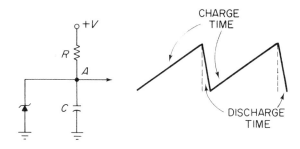

Figure 15-4. Zener Diode Sawtooth Generator

After a time, the charge on the capacitor becomes great enough to cause the Zener diode to go into avalanche breakdown. The avalanching Zener places a virtual short across the capacitor. Current from ground through the Zener—reverse current—discharges the capacitor very rapidly. The ramp voltage can be made quite linear by allowing the capacitor to charge to only 10 percent of the applied voltage. A capacitor charges to 10 percent of the applied voltage in about 0.08 time constant.

### 15.3.2 Selecting Component Values

There are so many possible combinations of $R$ and $C$ that decade trial components are recommended. Selecting the Zener is another matter. The firing potential of the Zener is the limit to the maximum amplitude of the output. The Zener must meet this requirement; otherwise its characteristics are not critical.

As one example a 200-k$\Omega$ resistor and a 30-$\mu$F capacitor might be used to produce a linear 480-ms ramp voltage. Time is calculated like this:

$$\begin{aligned} 1 \text{ time constant} &= RC \\ &= 200 \times 10^3 \times 30 \times 10^{-6} \\ &= 6 \text{ s} \quad \text{or} \quad 6000 \text{ ms} \end{aligned}$$

$$\begin{aligned} \text{ramp time} &= 0.08 \text{ time constant} \\ &= 0.08 \times 6000 \\ &= 480 \text{ ms} \end{aligned}$$

### 15.3.3 Synchronized Sawtooth Generator

The circuit of Figure 15-4 can be modified to a *synchronized sawtooth generator* by connecting a positive input trigger pulse to the top of the Zener diode. For proper operation, the frequency of the input trigger pulse must be just slightly higher than the free-run frequency of the oscillator. The Zener diode will then be triggered into avalanche conduction just an instant before it would do so voluntarily. Each input trigger will terminate one waveshape and start the next. The slope of the ramp voltage is still determined by the $RC$ time constant, but the frequency is the same as that of the trigger pulse.

## 15.4 Blocking Oscillator

The *blocking oscillator* produces a train of sharp trigger pulses. The repetition rate of these pulses is normally in the audio range. Once the circuit is established, the output frequency is stable. The timing pulses for this oscillator are narrow, with very steep rise and fall times.

### 15.4.1 Sample Circuit

A sample circuit of a blocking oscillator and waveshapes at base and output are illustrated in Figure 15-5. This is a free-running blocking oscillator that produces continuous outputs as long as the power is applied. When power is applied, there is zero bias on the transistor, but a small collector current starts through the primary of the transformer. This current causes a decrease in collector voltage which couples to point A on $L1$ as a positive potential. The transistor is now forward-biased and current increases. As current increases, feedback increases and causes further increases in current. The action is accumulative until the transistor saturates.

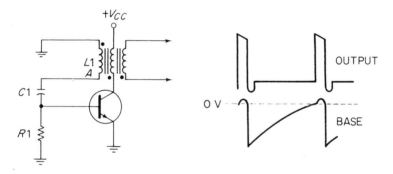

Figure 15-5. Blocking Oscillator

At saturation, the change in collector current ceases, and the collector voltage becomes momentarily steady. At this instant, feedback ceases. Meantime the capacitor has charged with negative on the base of the transistor. When feedback ceases, $C1$ starts to discharge through $R1$. This discharge current holds a negative potential on the base of the transistor. This negative potential is reverse bias. At the first instant, the reverse bias is very strong, and the transistor is driven from saturation to cutoff.

As the capacitor discharge cycle progresses, the reverse bias becomes smaller. The transistor remains cut off until the bias reaches zero. With a zero bias, the transistor starts to conduct again and initiates the next cycle.

The flattening of the top of the output signal is caused from collector saturation. The duration of the pulse at the top is the reaction time from saturation until reverse bias cuts off the transistor. The discharge of the capacitor is determined by the relative values of capacitance and resistance. This discharge time is the time from the termination of one pulse to the start of the next. So the values of $R$ and $C$ determine the output frequency of the oscillator.

### 15.4.2 Selecting Component Values

There is one critical feature to consider when selecting the transistor. It must be capable of handling the current when a voltage of $2 \times V_{CC}$ is applied to the collector. It must also be capable of oscillating at the desired frequency, but since the frequency is generally audio, the frequency problem is not serious.

Special blocking oscillator transformers are available, and they are recommended. However, almost any transformer can be used in this circuit.

The values of $R$ and $C$ are interdependent, and the best results can be obtained by experimenting with various values. Start with $R = 10$ k$\Omega$ and $C = 0.1$ $\mu$F and vary both components until the desired waveshape and frequency is attained. After that, fixed values can be substituted.

The capacitor requires one full time constant to discharge the required amount. At the suggested starting values for $R$ and $C$, the time between pulses is

$$t \approx RC$$
$$\approx 10 \times 10^3 \times 0.1 \times 10^{-6}$$
$$\approx 1 \text{ ms}$$

The frequency then is

$$f = \frac{1}{t} \approx \frac{1}{RC}$$
$$\approx \frac{1}{1 \times 10^{-3}}$$
$$\approx 1000 \text{ Hz}$$

The frequency can be reduced by increasing the value of either $R$ or $C$, or it can be raised by decreasing either of these values. Frequency can be made adjustable by using a variable component for either $R$ or $C$.

### 15.4.3 Synchronized Blocking Oscillator

A slight modification on the circuit of Figure 15-5 will change the free-running oscillator to a *synchronized blocking oscillator*. The only real difference is a fixed cutoff bias on the transistor. The transistor then remains cut off until it is triggered by a synchronizing pulse. Each time it is triggered, it produces one output pulse, then cuts off again. The circuit and waveshapes are shown in Figure 15-6. Notice that the only changes from Figure 15-5 are that the resistor is connected to $-V_{BB}$ instead of ground, and a synchronizing signal is applied to the base. The circuit in Figure 15-5 can be synchronized without adding the cutoff bias, but then the synchronizing frequency would have to be slightly higher than the free-running frequency. In either case, the output frequency is equal to the frequency of the synchronizing signal.

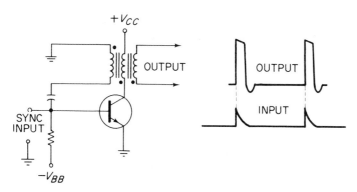

Figure 15-6. Synchronized Blocking Oscillator

## 15.5 Astable Multivibrator

The *astable multivibrator* is a free-running relaxation oscillator that produces a series of symmetrical square waves. It operates at a frequency determined by the circuit components and requires no signal input.

### 15.5.1 Sample Circuit

A sample circuit for an astable multivibrator is illustrated in Figure 15-7. The desirable output from an astable multivibrator is generally a symmetrical square wave. This is accomplished by matching the two

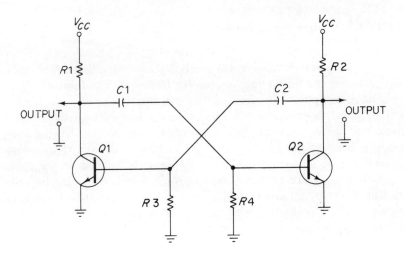

Figure 15-7. Astable Multivibrator

halves of the circuit as nearly as possible. However, a perfect balance is never quite achieved, so when power is applied, one transistor conducts more than the other. The voltage at the collector of the dominant transistor takes a sharp drop. The drop in potential forces the capacitor between the collector and the other base to start discharging. This reverse biases the slower transistor and cuts it off. The voltage at the collector of the cutoff transistor rises to $V_{CC}$ and causes the capacitor at that collector to start charging.

Suppose that $Q1$ is cut off and $Q2$ is saturated. Capacitor $C2$ is discharging through $R3$ and holding $Q1$ cut off. The collector voltage of $Q1$ is $V_{CC}$, and $C1$ is charging. Current through $R4$ to $C1$ holds forward bias on $Q2$ and keeps it saturated. When the voltage drop across $R2$ reaches about 0.1 V, $Q1$ starts to conduct. The drop in collector voltage of $Q1$ starts $C1$ discharging through $R4$. In a very short time, the situation has reversed, with $Q1$ saturated and $Q2$ cut off.

Two outputs are available, one from either collector. Both outputs are symmetrical square waves. At any given time, one of the outputs is positive and the other is negative.

### 15.5.2 Selecting Component Values

The transistors should be a matched pair capable of handling a full $V_{CC}$ collector voltage and saturation current. Since each transistor is saturated 50 percent of the time, heat dissipation may be a problem. Take the heat factor into consideration and use heat sinks if indicated. Fast switching capabilities are also of definite advantage.

The output frequency is largely determined by the *RC* time constants: $C1 \times R4$ and $C2 \times R3$. These two time constants should be equal. The frequency then is approximately equal to the reciprocal of one time constant.

$$f \approx \frac{1}{RC}$$

Only approximate values can be calculated in this fashion because the characteristics of the transistors affect the charge and discharge of the capacitors.

It is recommended that starting values of *R* and *C* be calculated, but the final values must be obtained by experimentation. For frequencies near 10 kHz, starting values should be

$$R3 \text{ and } R4 = 2 \text{ k}\Omega \text{ each}$$
$$C1 \text{ and } C2 = 0.05 \text{ }\mu\text{F each}$$
$$f \approx \frac{1}{RC}$$
$$\approx \frac{1}{2 \times 10^3 \times 0.05 \times 10^{-6}}$$
$$\approx 10 \text{ kHz}$$

The values of the two collector load resistors are not critical. They may be selected to match the impedance of the load they feed. If impedance matching is *not* a consideration, $R1$ and $R2$ should each be about 50 k$\Omega$. At any rate, $R1$ and $R2$ should be of equal value to maintain a symmetrical output.

## 15.6 Monostable Multivibrator

The *monostable multivibrator* is a useful circuit for delaying a signal for a specified period of time. It is also used to store information for very brief intervals. The monostable multivibrator is a synchronized relaxation oscillator. It has one stable state and must be triggered to force it out of this stable condition. When triggered, it produces one rectangular waveshape, then reverts to the stable state. The *RC* time constant largely determines the duration of the output waveshape, but the output frequency is the same as the triggering frequency.

### 15.6.1 Sample Circuit

A sample circuit for a monostable multivibrator is illustrated in Figure 15-8. Transistor $Q1$ is biased to cutoff by $-V_{BB}$ through $R5$ to its base. This is the stable state with $Q1$ cut off and $Q2$ saturated. In this condition, $V_{CC}$ on the collector of $Q1$ has charged $C1$ with a negative on the right-hand side. This condition holds until a positive trigger pulse is applied to the base of $Q1$.

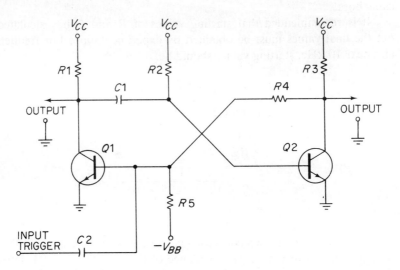

Figure 15-8. Monostable Multivibrator

The positive input trigger causes a surge of collector current through $Q1$ and a drop in collector voltage. Capacitor $C1$ starts to discharge through $R2$ and places reverse bias on $Q2$. The reverse bias cuts off $Q2$, and its collector voltage rises to $V_{CC}$. The high positive collector voltage on $Q2$ holds forward bias on $Q1$ and drives it to saturation. This is the unstable state during which a useful output is taken. This state will hold until the discharge of $C1$ allows $Q2$ to conduct again. The duration of the output is determined primarily by the $RC$ time constant of $R2 \times C1$. An output may be taken from either or both collectors. These two outputs are almost identical but 180 degrees out of phase. Output from $Q1$ is a negative-going wave while the output from $Q2$ is a positive-going wave.

### 15.6.2 Selecting Component Values

The transistors should be fast switching devices, and they must be capable of handling the current when full $V_{CC}$ is on the collector. The transistors need *not* be matched.

The resistor and capacitor values are not critical except for $R2$ and $C1$. The load resistors, $R1$ and $R3$, can be selected to match the impedance of the load they feed, but they should be approximately the same value. The feedback resistor, $R4$, should be about $10 \times R3$. Biasing resistor $R5$ should be about $2 \times R3$.

The value of $R2$ is generally about equal to that of $R4$; then a value of $C1$ is selected to produce a proper $RC$ time constant. The time constant should be about equal to the desired duration of the output. The starting values can be calculated, but the final values should be determined by bench testing the circuit for precise output waveshapes.

As trial values, set:

$R1$ and $R3$ to 10 kΩ each.
$R4$ to 120 kΩ.
$R5$ to 22 kΩ.
$R2$ to 120 kΩ.
$C1$ to 0.05 $\mu$F.

These values will produce an output of about 6 ms. This duration can be reduced by reducing the value of either $R2$ or $C1$. It can be increased by raising either of these values.

A faster switching time can be attained by placing a bypass capacitor across $R4$. This capacitor provides faster feedback and quicker saturation of $Q1$ after it is triggered to the on state. A slight alteration in the circuit of Figure 15-8 will enable the use of a single power supply. Simply connect the two emitters together and add a common resistor from emitters to ground. This resistor should be bypassed with a capacitor. When this is done, $R5$ can be disconnected from $-V_{BB}$ and returned to ground. These changes are illustrated in Figure 15-9.

## 15.7 Bistable Multivibrator

The *bistable multivibrator* is a very useful and popular synchronized oscillator. This circuit uses two transistors in a balanced state that provides two stable states. It requires an input trigger pulse to switch from

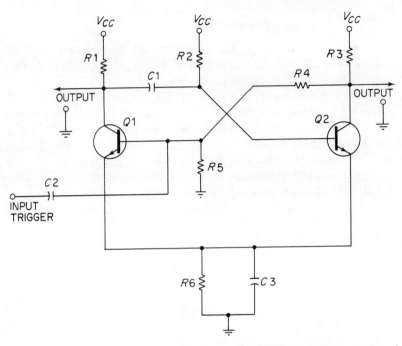

Figure 15-9. Monostable Multivibrator with a Single Power Supply

either stable state to the other. The square or rectangular wave outputs are used for a variety of timing and gating functions.

### 15.7.1 Sample Circuit

A sample circuit for a bistable multivibrator is illustrated in Figure 15-10. When power is applied, both transistors start to conduct, but this is a very brief condition. The dominant transistor quickly goes to saturation and holds the other cut off. This is a stable state that holds until a positive trigger pulse is applied to the base of the cutoff transistor. Each positive pulse input triggers a series of events that change the saturated transistor to cutoff and the cutoff transistor to saturation. The input pulse is applied to the base of both transistors, but it affects only the one that is cut off at that time. The bias supply voltage, $-V_{BB}$, is also applied to both bases. In conjunction with the feedback, the $-V_{BB}$ makes sure that only one transistor can conduct at any given time.

The output square waves can be taken from either or both collectors. The only difference between the two outputs is the phase relation; they are 180 degrees out of phase with each other. The frequency of these outputs is entirely controlled by the frequency of the input triggers. A

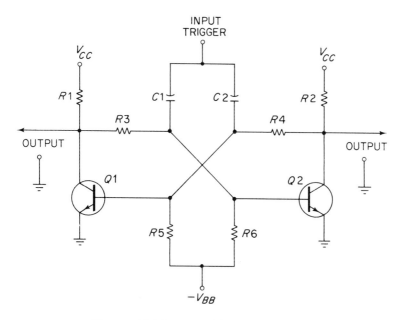

Figure 15-10. Bistable Multivibrator

half-cycle of each output is generated for each input. So the output frequency is half the frequency of the input.

### 15.7.2 Selecting Component Values

The two transistors should be as closely matched as possible, and they must have fast switching characteristics. The value of the resistors is not critical as long as the two halves of the circuit are perfectly balanced. The value of the two collector load resistors may be selected to match the load they feed. The other resistor values can then be selected in a certain relationship to these load resistors. For example: $R6$ should be about $3 \times R1$ and $R3$ should be about $3 \times R6$. This same relationship should be observed with $R2$, $R5$, and $R4$. Some adjustments will likely need to be made during testing, but good starting values are:

$$R1 \text{ and } R2 = 5.6 \text{ k}\Omega \text{ each}$$

$$R3 \text{ and } R4 = 68 \text{ k}\Omega \text{ each}$$

$$R5 \text{ and } R6 = 20 \text{ k}\Omega \text{ each}$$

The circuit as laid out in Figure 15-10 has an appreciable transit time. This time can be reduced by using bypass capacitors across $R3$ and $R4$. A suitable value for these capacitors is 150 pF each.

### 15.7.3 Steering Diodes

Maximum switching speed can be obtained with a given bistable multivibrator by applying the positive trigger pulses to only the cutoff transistor. This can be accomplished in the circuit of Figure 15-10 by adding two *steering diodes* to the input circuit. This modification is illustrated in Figure 15-11.

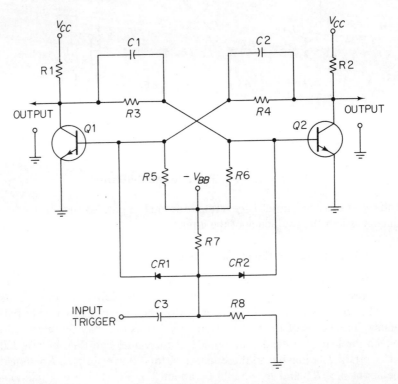

Figure 15-11. Bistable Multivibrator with Steering Diodes

The only additions to the circuit are bypass capacitors $C1$ and $C2$, resistors $R7$ and $R8$, and the two diodes, $CR1$ and $CR2$. Capacitors $C1$ and $C2$ increase the switching speed by providing very fast coupling for the feedback. The diodes increase the switching speed by steering the signal to the base of only the cutoff transistor. Resistors $R7$ and $R8$ provide diode bias to assist the steering action.

When $Q1$ is cut off, the next trigger pulse is applied to the base of $Q1$ only. When $Q1$ is cut off, the bias on $CR2$ is a very high reverse bias consisting of $V_{CC}$ on the cathode and $-V_{BB}$ on the anode. The bias on $CR1$ at this time is close to zero; it has a small negative potential on the anode and nearly zero on the cathode. The next positive input pulse through $C3$ causes $CR1$ to conduct and couple the trigger pulse to the base of $Q1$.

When $Q2$ is cut off, the next trigger pulse is applied to the base of $Q2$ only. When $Q2$ is cut off, the bias on $CR1$ is a very high reverse bias consisting of $V_{CC}$ on the cathode and $-V_{BB}$ on the anode. The bias on $CR2$ at this time is close to zero; it has a small negative potential on the anode and nearly zero on the cathode. The next positive input through $C3$ causes $CR2$ to conduct and couple the trigger pulse to the base of $Q2$.

The diodes should be signal diodes with reverse voltage capabilities equal to about twice the value of $V_{CC} + V_{BB}$.

SECTION SIX

# POWER SUPPLIES

*Power-supply circuits* must be designed to convert the commercially available ac to usable dc. This task requires rectifiers, filters, voltage multipliers, current regulators, voltage regulators, and voltage dividers. For purpose of discussion, these circuits are divided into two types: conversion circuits and regulator circuits.

CHAPTER SIXTEEN

# Conversion Circuits

*Power conversion circuits* include rectifiers, filters, multipliers, dividers, and converters. These circuits convert ac to dc, remove the ripple, set dc to the desired level, and, in some cases, change dc back to ac. Some form of energy-conversion circuit is required in nearly all electronic equipment. The most convenient form of delivering electrical energy to the user is alternating voltage, and in the United States it is at a frequency of 60 Hz. If any part of the designed equipment requires anything except a 60-Hz alternating voltage, the designer must convert this voltage to fit the particular need.

## 16.1 Power Transformer

A power supply for solid-state equipment must deliver a relatively high current with a low voltage. These characteristics must be considered in the design of the power supply. The *transformer* in the power supply must be designed specifically for use with solid-state circuits. This type of transformer will have relatively few turns on the secondary, to enable it to deliver a stepped-down voltage. At the same time the cross-sectional area of the wire must be large in order to carry the high current that is needed. This large wire reduces the resistance and keeps losses to a minimum.

Power frequencies range from 60 to 1600 Hz, but a particular transformer is designed to handle only one frequency within this range.

### 16.1.1 Transformer Efficiency

A perfect transformer has a 1:1 ratio between the input power to its primary and the output power that it delivers from its secondary.

There are no perfect transformers, but some transformers do deliver output power that is more than 99 percent of the input power. The efficiency of a transformer is calculated by dividing the power out by the power in. This efficiency is normally given in percentage, so 100 times the result is efficiency in percentage.

$$\% \text{ efficiency} = \frac{P_{out}}{P_{in}} \times 100$$

A certain power transformer has a primary power of 48 W and a secondary power of 47 W, as illustrated in Figure 16-1.

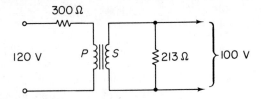

Figure 16-1. Power Transformer

This particular transformer has an input power of

$$P = \frac{E^2}{R}$$
$$= \frac{14,400}{300}$$
$$= 48 \text{ W}$$

The output power is

$$P = \frac{E^2}{R}$$
$$= \frac{10,000}{213}$$
$$= 47 \text{ W}$$

The efficiency of this transformer is

$$\% \text{ efficiency} = \frac{P_{out}}{P_{in}} \times 100$$

# Conversion Circuits

$$= \frac{47}{48}$$
$$= 97.9\%$$

The 1 W of power that is lost between primary and secondary is an energy loss due to copper losses, hystersis losses, and eddy-current losses.

### 16.1.2 Transformer Ratios

Normally 120 V ac is delivered to the power line. The power transformer must step this 120 V down to a usable level. It is an easy matter to design a transformer that will step the voltage down while stepping the current up. It is a matter of using a primary with more turns than the secondary. The relationship between turns and voltage is

$$\frac{E_p}{E_s} = \frac{N_p}{N_s}$$

where $E_p$ is voltage across the primary and $E_s$ is voltage across the secondary. The actual number of turns is not important; it is the ratio that counts.

The impedance of the primary is also dependent on the turns ratio, since part of the secondary impedance is reflected into the primary.

$$\frac{Z_p}{Z_s} = \left(\frac{N_p}{N_s}\right)^2$$

### 16.1.3 Phase Relationship

In a single-phase power supply, the output must be either in phase or 180 degrees out of phase with the input. In phase or out of phase is determined by how the secondary winding is connected.

## 16.2 Power Rectifiers

An ideal *power rectifier* has zero resistance in the forward direction and an infinite resistance in the reverse direction. In practice, this ideal is *not* available. The best available rectifier for a particular power supply can be selected only after considering the operating conditions that it will be exposed to.

### 16.2.1 Resistance Ratio

Normally a special high-priced rectifier is *not* required for a solid-state power supply. If voltage and current are kept within specified limits and the diodes are not subjected to abuse, ordinary diodes are satisfactory. These diodes have a very small forward resistance. The reverse resistance of these is very large in comparison to the load resistance.

### 16.2.2 Peak Inverse Voltage

Sometimes the peak inverse voltage rating of a diode is less than the value needed for a power supply. This condition can be corrected by connecting two diodes in series. This doubles the PIV capacity, but it also doubles the forward resistance. The increase in forward resistance decreases the output-voltage amplitude and increases the operating temperature. If the temperature is likely to approach the rated value, it is best to use diodes with heat sinks to help dissipate the extra heat.

### 16.2.3 Increasing the Current Capacity

When the required current exceeds the capacity of a single diode, two diodes may be connected in parallel. This parallel arrangement doubles the current capacity. If parallel diodes are used, they must be carefully matched. Otherwise, one diode will have to do most of the work.

### 16.2.4 Transient Considerations

Opening and closing switches tend to cause surges of voltage and current that are much greater than normal amplitudes. If a power supply is subject to reasonably high surges it may be advisable to choose germanium or selenium rectifying elements instead of silicon diodes. A good silicon diode rectifier has extremely low reverse current until it reaches the avalanche breakdown stage. When a rectifying diode breaks down, it is destroyed. Germanium and selenium rectifiers have more reverse leakage current, and this gives them a better chance to absorb the surges without damage.

## 16.3 Half-wave Rectifier

Many solid-state circuits require a relatively high voltage with very little current. The *half-wave rectifier* is well suited for supplying such circuits, yet it is the simplest of all power-supply circuits.

### 16.3.1 Sample Circuit

Figure 16-2 is a sample circuit for a half-wave rectifier with unfiltered output. This diode conducts each time that its anode goes positive. The output is a string of positive pulses that closely resemble the positive alternation of each sine wave on the secondary of $T1$. When the anode goes negative, the diode cuts off, and, except for a slight leakage current, there is no output. A negative output can be obtained by reversing the connections on the diode.

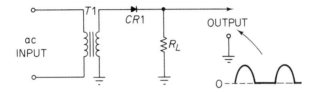

Figure 16-2. Half-wave Rectifier

### 16.3.2 Selecting Component Values

The diode in a half-wave rectifier is subjected to a high peak inverse voltage. This fact must be considered when selecting the diodes. When the diode is cut off, the voltage across it is equal to the peak-to-peak value of the voltage across the secondary of the transformer. This amplitude is approximately 2.83 times the RMS value of the secondary voltage.

The peak amplitude of the pulsating dc output is equal to the peak amplitude of the secondary voltage except for the slight voltage drop across the conducting diode. This difference is so small that it can be ignored. The average value of voltage during one alternation of a sine wave is $0.637 \times$ peak value. Since the half-wave rectifier uses only every second alternation, its average output is $0.318 \times$ peak amplitude.

Resistor $R_L$ is a bleeder resistor to complete the circuit when all external loads are turned off. Its value should be high enough to allow only a trickle of current. A value of 1 to 5 M$\Omega$ is recommended.

The hazard to the diode due to current surges can be greatly reduced by adding a small value of resistance in series with the diode. This resistor should be placed between the transformer winding and the diode. Many circuit and load conditions affect the amplitude of current surges. The series resistance value must be selected experimentally under normal operating conditions. The resistance value should be as low as possible, but must be large enough to dampen all surges to tolerable values.

## 16.4 Full-wave Rectifier

The *full-wave rectifier* is well suited for supplying circuits that require relatively low voltage and high currents. The output is easy to filter and regulate and can be used with nearly all solid-state circuits.

### 16.4.1 Sample Circuit

A sample circuit and unfiltered output of a full-wave rectifier are illustrated in Figure 16-3. Each diode conducts during the portion of the cycle that causes its anode to become positive. The current path for each diode is from ground, through $R_L$, through the diode, and out of the center tap of the transformer back to ground. Each alternation of the input develops a positive pulse across $R_L$.

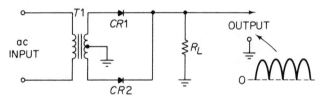

Figure 16-3. Full-wave Rectifier

The peak amplitude of each output pulse is approximately half the peak value of the secondary voltage. The ripple frequency of the pulsating dc is twice the frequency of the ac input. When a diode is cut off, it has a peak inverse voltage equal to half the peak amplitude of the secondary voltage.

### 16.4.2 Selecting Component Values

The full-wave rectifier is less vulnerable to current surges, and the inverse peak voltage across the diodes is much less than in a half-wave rectifier. The main consideration in selecting the diodes should be maximum reverse voltage, forward voltage drop, leakage current, and forward current. The average reverse voltage rating on the data sheet should be about 2 × peak output. The forward voltage drop is of little concern except as it subtracts from the output amplitude. The smaller the forward voltage drop, the greater the output amplitude.

Conversion Circuits

Reverse leakage current when the diode is cut off is more of an asset than a disadvantage. At most, the reverse current is only a few milliamperes, and this reverse current helps to absorb current surges in the power supply. It is recommended that diodes with relatively large reverse currents be utilized in a full-wave rectifier.

The maximum forward current rating must be about 2 × maximum output current. This is a safety margin over the quantity of current that the diode must carry at any instant.

The bleeder resistor, $R_L$, should be a large value on the order of 1 to 5 M$\Omega$.

## 16.5 Bridge Rectifier

The *bridge rectifier* is the most suitable all-purpose rectifier for solid-state power supplies. It incorporates the high current characteristics of the full-wave rectifier with the high voltage of the half-wave rectifier. The circuit is simple and economical and is recommended for all types of common solid-state circuits.

### 16.5.1 Sample Circuit

A sample bridge rectifier circuit and its unfiltered output are illustrated in Figure 16-4. On one alternation the current path is from ground, through $R_L$, through $CR2$, through the transformer, and through $CR1$ to ground. On the next alternation the current path is from ground, through $CR4$, through the transformer, and through $CR3$ to ground. The unfiltered output is a full-wave rectification of the ac sine wave with an amplitude approximately equal to the peak value of the voltage

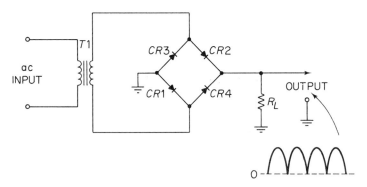

Figure 16-4. Bridge Rectifier

across the secondary of T1. The frequency of the ripple on the pulsating dc output is twice the frequency of the ac input. This output is easily filtered and provides both high current and high voltage.

### 16.5.2 Selecting Component Values

The diodes for the bridge rectifier should meet the same criteria as described for full-wave rectifiers, and $R_L$ should have a large value of resistance. The peak output across $R_L$, added to the forward voltage drop across two diodes in series, is equal to the peak voltage across the secondary of $T1$.

It is possible to obtain four diodes already connected in a bridge configuration. If such a package is used, consider the data-sheet ratings as applying to each individual diode.

## 16.6 Power Supply Filters

As a rule, elaborate filtration is not needed for solid-state power supplies. Solid-state circuits in general are not greatly affected by a small ripple on the direct voltage. Where voltage levels are critical, a good voltage regulator is more important than perfect filtration.

### 16.6.1 Capacitor Input Filter

The output of each of the rectifiers previously discussed can be adequately filtered by a capacitor input filter. In each case, the filter is complete with the addition of a single capacitor. The capacitor is connected in parallel with the bleeder resistor, as indicated in Figure 16-5.

This is the same bleeder resistor shown in all three rectifier circuits. The discharge path of the capacitor in this drawing is true only when the filtered dc output is *not* being used. This resistor, $R_L$, has a resistance in excess of 1 MΩ while the effective resistance of the external load may

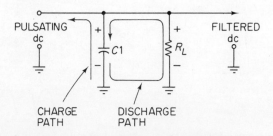

Figure 16-5. Capacitor Input Filter

be as low as 2 or 3 Ω. Obviously under these conditions, the filter consists of $C1$ and the effective resistance of the external load.

The drawing in Figure 16-5 points up an important use of the bleeder resistor, $R_L$. When the external load is disconnected, $R_L$ offers the only discharge path for $C1$. Without this bleeder, capacitor $C1$ can hold a dangerously high charge for several days.

### 16.6.2 Selecting Capacitor C1

There are two criteria for selecting a filter capacitor: dc voltage rating and proper capacitance. The voltage rating is based on the output voltage level. A voltage rating equal to 1.5 times the output voltage is recommended.

The capacitance of $C1$ must be based on the resistance of the external load, since this is the filter resistance. The capacitor should be an electrolytic type because of the high capacitance required. Another point to consider is the percentage of ripple that can be tolerated. When these factors have been considered, the capacitance of $C1$ in microfarads is

$$C1 = \frac{200 \times 10^3}{R \times \text{ripple}}$$

where $R$ is the resistance of the external load and the percentage of ripple is expressed as a whole number. For example, when output voltage is 30 V, external load resistance is 6 Ω, and the tolerable ripple is 5 percent, the capacitance is

$$C1 = \frac{200 \times 10^3}{R \times \text{ripple}}$$
$$= \frac{200 \times 10^3}{6 \times 5}$$
$$= 6667 \ \mu F$$

The resistance of the external load is easily determined by measuring the output voltage and current. Determining the resistance is a simple application of Ohm's law: $R = E/I$.

## 16.7 Voltage Multipliers

In some cases, solid-state circuits require extremely high voltages with very little current. *Voltage multipliers* can be used to raise the direct voltage level to any reasonable value. These multipliers are voltage doublers, voltage triplers, and voltage quadruplers. However, the quadrupler is a high-loss circuit, and it should be replaced when the need arises by two voltage doublers. When a multiplier is used in conjunction with a rectifier, no further filtering is required. Devices that use very high voltage seldom use any appreciable current. This low current enables the use of multipliers without worrying about filtration.

### 16.7.1  Voltage Doubler

A sample circuit for a cascade voltage doubler is illustrated in Figure 16-6. When point A of $T1$ goes negative, electrons leave this point and flow into $C1$. From $C1$ the electrons flow through $CR2$ back to the positive side of the transformer winding. This action charges $C1$ to the peak value of the voltage on the secondary of $T1$.

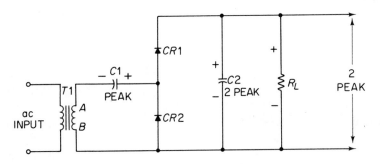

Figure 16-6.  Cascade Voltage Doubler

On the next alternation of the input, point B on the transformer goes negative. The charge on $C1$ is in series, aiding with the secondary voltage. Electrons from $C1$ flow into the transformer winding, and electrons from point B flow into $C2$. From $C2$, electrons flow through $CR1$ back to $C1$. This action places a charge on $C2$ that is equal to twice the peak value of the voltage across the secondary of $T1$.

The values of $C2$ and $R_L$ form a very long time constant in respect to the 60-Hz input frequency. As a result of this long time constant, $C2$ loses practically no charge during its noncharging time. The positive output across $R_L$ is equal to the charge on $C2$: twice the peak of the voltage across the secondary of $T1$. For best results, the external load that is connected in parallel with $R_L$ must be an extremely large resistance; at least $10 \times R_L$. A small resistive load will draw current, discharge $C2$, and defeat the purpose of the voltage doubler.

## 16.7.2 Selecting Component Values

Each of the diodes for this circuit must have a peak inverse voltage rating in excess of twice the peak value of the secondary voltage. A minimum leakage current and a minimum forward voltage drop are also assets to be considered. Each time power is applied, the diodes are subjected to high levels of current. This is a brief period of time, and heating is not a problem; but make sure that the diodes can handle this heavy current without damage.

Both $C1$ and $C2$ should be electrolytic capacitors because their value must be quite large. The working voltage of $C1$ must be in excess of twice the secondary peak voltage, and the working voltage of $C2$ must be twice that of $C1$.

The time constant of $R_L \times C2$ must be at least 10 times the duration of a 60-Hz sine wave. A time constant many times this value is recommended. In fact, an $R_L$ of 5 M$\Omega$ and a $C1$ of 50 $\mu$F are *not* extravagant. The capacitance of $C1$ should be about 25 $\mu$F.

## 16.7.3 Voltage Tripler

The cascade voltage doubler is easily converted to a voltage tripler by adding another section, consisting of a rectifier, a resistor, and a capacitor. The tripler circuit is represented in Figure 16-7. The voltage-doubler portion of the circuit has *not* been disturbed. It still has the same components and performs voltage doubling exactly as previously described. The added section, consisting of $CR3$, $C3$, and $R_L'$, simply

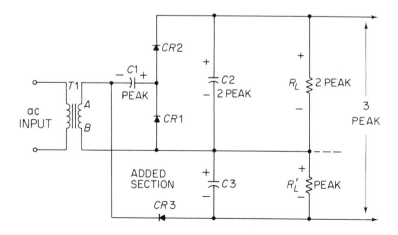

Figure 16-7. Voltage Tripler

adds another peak value to the already doubled voltage. The output is now equal to three times the peak value of the voltage across the secondary of $T1$.

In this circuit, when point A on the transformer goes negative, $C1$ conducts to charge $C1$; and $CR3$ conducts to charge $C3$. On the next alternation $C2$ charges to $2 \times$ peak, as previously described. The output is taken across the combination of $C2R_L$ and $C3R_L'$, and this is $3 \times$ peak.

Rectifier $CR3$ should have characteristics similar to those of $CR1$ and $CR2$. Capacitor $C3$ should be the same as $C1$, and the value of $R_L'$ should be half that of $R_L$.

Remember, voltage multipliers will *not* work properly with a load that draws any appreciable current. The instant that current is drawn, the output capacitors tend to discharge, and the output voltage level takes a drastic drop.

### 16.8 Voltage Dividers

*Voltage dividers* are simple but essential circuits in nearly all electronic power supplies. Without voltage dividers, the power supply would need a separate rectifier circuit for each different level of voltage used for the various loads. These voltage dividers range from a simple divider with a single fixed load to complex dividers with multiple changing loads.

#### 16.8.1 Divider for a Single Fixed Load

This is an easily constructed circuit, and it is needed constantly for biasing purposes. The layout is similar to that in Figure 16-8. The voltage divider consists of $R1$ and $R2$. The load, represented by $R_L$, must

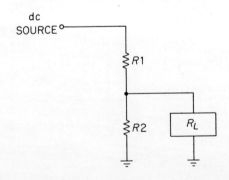

Figure 16-8. Single Load Divider

# Conversion Circuits

have a specified voltage across it in order to produce the required current. The parallel combination of $R2$ and $R_L$ must produce enough current to drop the source voltage *less* the voltage required across the load.

Suppose that the load resistance is 2 k$\Omega$ and requires a current of 3 mA. This means that 5 V must be delivered to the load for proper operation. Further suppose that the source voltage is $+20$ V. This means that $R1$ must drop 15 V in order to place exactly 5 V across the load.

Using a 3-k$\Omega$ resistor for $R1$, the 3 mA from $R_L$ causes a voltage drop across $R1$ of

$$E = IR$$
$$= 3 \times 10^{-3} \times 3 \times 10^3$$
$$= 9 \text{ V}$$

This leaves 20 V $-$ 9 V, or 11 V, and this is too much voltage for the load. The voltage drop across $R1$ must be increased by another 6 V. This means that $R2$ must produce 2 mA of current to drop this additional 6 V across $R1$. Since $R2$ is in parallel with $R_L$, it has the same 5 V across it that there is across $R_L$. What value of resistance will produce 2 mA of current with a voltage of 5 V?

$$R = \frac{E}{I}$$
$$= \frac{5 \text{ V}}{2 \text{ mA}}$$
$$= 2.5 \text{ k}\Omega$$

When $R2$ is given a value of 2.5 k$\Omega$, there are 2 mA through $R2$, 3 mA through $R_L$, and 5 mA through $R1$. The voltage drop across $R1$ is

$$E = IR$$
$$= 5 \times 10^{-3} \times 3 \times 10^3$$
$$= 15 \text{ V}$$

This leaves the required 5 V to appear across the load resistance and across $R2$.

In practice it is recommended that $R2$ be a variable resistor during the bench testing. The value of $R2$ can then be varied until the proper load voltage is present; then a fixed resistor can be substituted for $R2$.

When the load requires a variable voltage, R1 and R2 can be replaced with a potentiometer with $R_L$ connected to the variable arm. A variable load current can be provided by connecting a rheostat in series with $R_L$.

### 16.8.2 Multiple Load Divider

The more circuits that are powered by a voltage divider, the more complex the divider becomes. As illustrated with the single load divider in Figure 16-8, a voltage divider is a series–parallel circuit. Each load that is added or removed from the divider causes a redistribution of voltage which affects all other loads. Figure 16-9 illustrates a voltage divider that is supplying voltage to three different circuits.

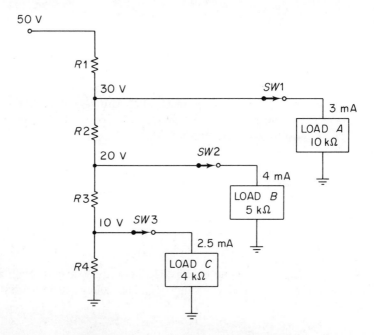

Figure 16-9. Multiple Load Divider

The current times the resistance of each load indicates the voltage requirement for each load. The values of R1, R2, R3, and R4 must be selected so that the divider furnishes 30 V to load A, 20 V to load B, and 10 V to load C. This means that R1 must drop 20 V, R2 must drop 10 V, R3 must drop 10 V, and R4 must drop 10 V.

Since R4 is in parallel with load C of 4 kΩ, let the resistance of R4 match that of load C. This sets $R4$ to 4 k$\Omega$, and the 10 V across it produces a current of 2.5 mA.

Resistor $R3$ carries the current of $R4$ plus the current of load C. This 5 mA through $R3$ must drop the 10-V difference between load B voltage and load C voltage:

$$R3 = \frac{E}{I}$$

$$= \frac{10 \text{ V}}{5 \text{ mA}}$$

$$= 2 \text{ k}\Omega$$

Resistor $R2$ carries the 5 mA of $R3$ plus the 4 mA from load B. This 9 mA of current through $R3$ must drop the 10-V difference between load B voltage and load A voltage:

$$R2 = \frac{E}{I}$$

$$= \frac{10 \text{ V}}{9 \text{ mA}}$$

$$= 1.1 \text{ k}\Omega$$

Resistor $R1$ carries the 9 mA from $R2$ and the 3 mA from load A. This 12 mA must drop 20 V across $R1$: the difference between the 30 V needed and the 50 V of the source:

$$R1 = \frac{E}{I}$$

$$= \frac{20 \text{ V}}{12 \text{ mA}}$$

$$= 1.67 \text{ k}\Omega$$

The wattage rating of each resistor must be at least twice the required power ($I^2R$).

The combination $R1$, $R2$, $R3$, and $R4$ may be obtained as one multiple-tapped, wire-wound resistor. The taps can then be adjusted to provide the proper voltage to each load.

It should be emphasized that the ideal voltage across each load is possible only when all loads are connected and functioning properly. Open switch 2 and calculate the results on the remaining loads.

With all switches closed, the total resistance of the circuit (divider and loads) is 4.17 kΩ. Removing the 5 kΩ of load B from the circuit increases the total resistance to slightly over 5 kΩ. The total current is now 10 mA instead of 12 mA. About 17 V is dropped across $R1$, and this leaves load A with 33 V instead of the desired 30 V. Current through load A goes up to 3.3 mA instead of the desired 3 mA. This is a 10 percent increase in current, and probably is within tolerance.

What about load C? The current through $R2$ and $R3$ is 10 mA − 3.3 mA, or 6.7 mA. This drops 7.37 V across $R2$ and 13.4 V across $R3$. Subtracting these voltages from the 33 V across load A leaves about 12.2 V across load C. The current of load C increases from 2.5 mA to slightly over 3 mA. This is more than a 20 percent increase in current, and in many cases, this would cause problems. The discrepancy of course grows worse as more switches are opened.

During bench testing of the voltage divider, it should be determined exactly what effect the removal of each load will have on the remaining loads. When all loads are removed except one, the voltage and current for the remaining load must be within tolerance. This is a tedious process but a necessary one. The loads can be simulated and measurements taken under all possible conditions. Usually the ideal starting voltages must be sacrificed in order to be within tolerance at all times.

CHAPTER SEVENTEEN

# Voltage Regulators

In many cases, it is necessary to have a direct voltage or current that remains at a steady level (within tolerance) in spite of changes in input voltage and circuit conditions. This is *not* ripple voltage. If ripple voltage is a problem, it can be solved by using a better filter. The problem that requires a *voltage regulator* is an actual shift in the level of the direct voltage. The output voltage from a rectifier circuit will change its level when the

Input voltage changes.
Load resistance changes.

There are several types of regulators to solve this voltage problem.

## 17.1  Zener Diode Regulator

The *Zener diode voltage regulator* is probably the simplest form of voltage regulator, yet it does an adequate job of regulating voltage for many solid-state requirements. The Zener characteristics, capabilities, and limitations must be considered, but, within limits, it is a good regulator.

### 17.1.1  Sample Circuit

Figure 17-1 illustrates a Zener diode voltage regulator. This is a shunt regulator, and the load resistance is connected in parallel with the Zener diode. The Zener diode is operating in the center of the Zener region. This provides a specified level of reverse current through the

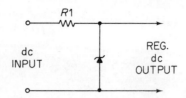

Figure 17-1. Zener Diode Regulator

Zener in the direction of the arrow. The desired voltage across the load is equal to the Zener operating voltage. As long as the Zener limits are not exceeded, the regulated voltage output will be maintained at the Zener voltage within a tolerance of $\pm 0.4$ V.

An increase of input voltage, or an increase in load resistance, tends to cause an increase in the voltage across the Zener and the load. Instead of a voltage increase, the Zener current increases. The extra current causes more voltage drop across the series resistor, $R1$, and this absorbs the original change. The output dc holds at a steady level.

A decrease of input voltage, or a decrease in load resistance, causes a tendency to reduce the voltage across the load and across the Zener. Instead of a voltage decrease, the Zener current decreases. The smaller current causes less voltage drop across $R1$ and cancels the original change. The output regulated dc holds at a steady level.

### 17.1.2 Selecting Component Values

The critical design characteristics for the Zener diode are avalanche voltage and maximum current and power capability. The maximum power is avalanche (Zener) voltage times the maximum current rating. Zener diodes are available with a wide variety of Zener voltage and maximum current ratings. Choose one with a Zener voltage equal to the desired regulated voltage level.

The value of $R1$ must be calculated in terms of minimum and maximum circuit currents to keep the Zener functioning in the Zener region (a Zener current between zero and the rated maximum). In order to do this, the minimum and maximum values of the input voltage must be considered.

### 17.1.3 Design Example

Problem: Design a Zener regulator to regulate the voltage across a load to 30 V. The input voltage is subject to change from 40 V to 60 V.

The Zener required for this circuit must have a 30-V breakdown (Zener) voltage and a maximum current rating in excess of 15 mA.

Voltage Regulators

The series resistor, $R1$, must prevent zero Zener current at 40 V input and allow no more than 15 mA with a 60-V input. The voltage drop across $R1$ may range from 10 V to 30 V. The proper value for $R1$ is

$$R1 = \frac{V_{max}}{I_{max}}$$
$$= \frac{30 \text{ V}}{15 \text{ mA}}$$
$$= 2 \text{ k}\Omega$$

The extreme conditions are illustrated in Figure 17-2. In both cases, the calculations are assuming an infinite load resistance. Adding any load at all will reduce the Zener current by an amount equal to the load current:

$$I_L = \frac{30 \text{ V}}{R_L}$$

At the two input voltages specified, the current through $R1$ will not change. All changes of load current are absorbed by changes in Zener current.

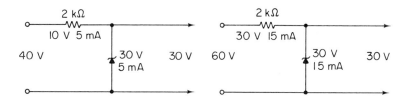

Figure 17-2. Extreme Regulating Conditions

## 17.2 Electronic Shunt Regulator

The *electronic shunt voltage regulator* is capable of regulation over a wider range of current changes than the Zener regulator just discussed. Of course, the current capacity of the Zener regulator could be increased by placing several Zener diodes in parallel, but this arrangement creates additional problems. It is best to use a Zener diode for bias control and have transistors handle the heavy current.

### 17.2.1 Sample Circuit

A sample circuit for an electronic shunt voltage regulator is illustrated in Figure 17-3. The Zener diode, $CR1$, operates in the breakdown (Zener) voltage region and maintains its reference voltage drop within the limits of zero to maximum Zener current. If input voltage or load resistance increases, the output voltage tends to increase. This action causes an increase in Zener current. Zener current must pass through $R2$. So the current increase causes more voltage drop across $R2$, and the base potential of $Q2$ goes more positive, increasing the forward bias. The emitter current of $Q2$ increases, and this increases the forward bias on $Q1$. The current through $Q1$ increases, causing more current through $R1$. The additional voltage drop across $R1$ offsets the original increase and holds the output to a steady value.

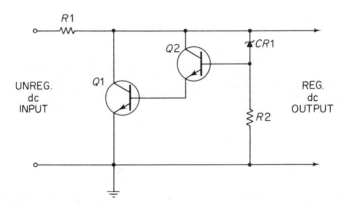

Figure 17-3. Electronic Shunt Regulator

A decrease in input voltage, or a decrease in load resistance, tends to decrease the output voltage. Either of these changes causes a decrease in Zener current. The smaller current through $R2$ reduces the potential on the base of $Q2$. The reduced forward bias on $Q2$ causes the emitter current to decrease, and this decreases the forward bias on $Q1$. The current through $Q1$ decreases and causes less current through $R1$. The smaller current through $R1$ drops less voltage, and this offsets the original decrease. The output voltage is held to a steady level.

A properly designed regulator of this type will maintain the regulated output level to a constant value within a tolerance of $\pm 0.5$ percent.

### 17.2.2 Design Example

Assume a dc power supply with a voltage level that is subject to vary from 45 V to 55 V. The load requires a voltage of 28 V, and this must be closely regulated. This regulation can be adequately handled by an electronic shunt regulator as represented in Figure 17-3.

The main considerations for the transistors are that they be free of noise and have a high current capacity. Under no load conditions the current through $Q1$ may approach 1 A. Cooling fins may be required.

The voltage drop across $R1$ will vary with the input, and it will range from 17 V to 27 V. A 28-$\Omega$ resistor should be adequate for this purpose. In some cases, $R1$ will be required to dissipate a power of as much as 26 W. The wattage rating for this resistor should be at least 50 W.

The Zener diode should have a reference voltage just slightly less than the desired regulated output. In this case, the desired output is 28 V, so the Zener should have a reference voltage of 27 V for currents between zero and 0.5 A.

The biasing resistor $R2$ should supply a voltage of $+1$ V on the base of $Q2$ when the Zener current is on the order of 1 mA. In fact, the average voltage drop across $R2$ must be kept at 1 V in order to keep a 28-V output. Under the most adverse conditions, the current through $R2$ should not exceed 16 mA; so a 0.5-W resistor should be more than adequate.

The shunt regulator does an adequate job of regulating voltage, but it does this at a cost of wasted power. The regulating component is in parallel with the load, and the regulating current is not available to the load. In addition, the shunt regulator places a continuous load on the power supply. These disadvantages should be considered in deciding about the advisability of using a shunt regulator.

## 17.3 Basic Series Regulator

The regulating device in a *series regulator* is placed in series with the load resistance, and the load current becomes the regulating current. This is a more economical use of power, and when the load is turned off, the drain on the power supply is negligible. There are many varieties of series regulators to choose from; they range from very basic to quite elaborate in design.

### 17.3.1 Sample Circuit

Figure 17-4 is a sample circuit for the most basic type of series voltage regulator. Transistor $Q1$ is the regulating device, and all load current passes through this transistor. The transistor acts as a variable resistance and adjusts its internal resistance to offset any reasonable change in either input voltage or load resistance.

Resistor $R1$ and $CR1$ control the forward bias on $Q1$ to regulate its internal resistance. A fixed voltage is held across the Zener diode, and all changes in Zener current cause changes in voltage across $R1$. Changing voltage across $R1$ varies the bias on $Q1$.

An increase in the input voltage, or an increase in the load resistance, tends to cause an increase in the output voltage. Either of these actions causes an increase in Zener current, and more voltage is dropped across $R1$. This increase in voltage across $R1$ decreases the forward bias on $Q1$ and increases its internal resistance. The voltage drop across $Q1$ increases to offset the original increase. The output voltage remains at a steady level.

A decrease of the input voltage, or a decrease in the load resistance, tends to decrease the output voltage. Either of these actions causes Zener current to decrease and reduce the voltage drop across $R1$. The smaller voltage across $R1$ increases the forward bias on $Q1$ and decreases its internal resistance. The voltage drop across $Q1$ decreases and offsets the original decrease. The output voltage is held to a steady level.

### 17.3.2 Design Example

Assume a power supply with an output that is subject to vary from 9 V to 14 V. The voltage for the load must be regulated to 7 V within a tolerance of ±0.5 V. Normally the load current will *not* exceed 4 A. This is a simple regulating problem that can be handled by a series regulator such as that in Figure 17-4.

The transistor will be exposed to rather high current. It should be a power transistor with a heat sink mounting. This transistor must be able to handle a collector voltage of 30 V with a current of 4 A and still have a reasonable margin for safety. The transistor will be required to drop 2 V with the minimum input and 7 V with the maximum input.

The Zener diode should have a breakdown voltage very near the required regulated voltage level. In this case, a 6.5-V 0.5-A Zener should do the job. The Zener power rating should be at least 0.1 × maximum output power. The output load current should never exceed 10 times the maximum Zener current.

## Voltage Regulators

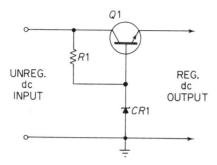

Figure 17-4. Basic Series Regulator

Resistor $R1$ must have a value that will properly bias the transistor for all values of Zener current. When the input voltage is at a minimum, the current through $R1$ should be about 5 percent of the output load current. The voltage across the Zener is 6.5 V; so with 9 V input, the voltage across $R1$ is 9 V − 6.5 V = 2.5 V. The maximum load current is 4 A, and 5 percent of 4 A = 0.1 A.

$$R1 = \frac{2.5 \text{ V}}{0.1 \text{ A}}$$
$$= 25 \ \Omega$$

This is a trial value for $R1$, and it may require some adjustment during bench testing.

### 17.3.3 Slowing the Action

The basic regulator in Figure 17-4 acts extremely fast to correct any tendency toward change. A small surge of voltage or current will cause an unnecessary correction. This tends to keep the regulator in a constant state of flux. The action needs to be slowed down so that corrections are made only when the change indicates a definite shift that needs to be corrected. A capacitor placed in parallel with the Zener diode as indicated in Figure 17-5 will slow down the corrective action. The capacitor, $C1$, should be an electrolytic capable of handling voltages slightly in excess of the Zener voltage. The capacitance for this circuit should be about 30 µF. The capacitor charges to the Zener voltage. Fast transients are absorbed by the capacitor and cause very little voltage change across $R1$. The slower changes allow time for the change to register on $R1$; then corrective action is taken.

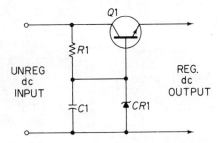

Figure 17-5. Slowing the Action

## 17.4 Electronic Series Regulator

The series voltage regulator can be greatly improved by adding a few more components. Some of the improvements that can be made are greater stability over a wider range of current and voltage and closer tolerance in the regulated output.

### 17.4.1 Sample Circuit

Figure 17-6 is a sample circuit for an electronic series voltage regulator with a fixed output voltage. Transistor $Q1$ is the series control device or regulator transistor. Its bias determines the internal resistance, conduction level, and series voltage drop across it. Transistor $Q2$ is the bias control for $Q1$. The current through $Q2$ determines the voltage drop across $R1$ which varies the potential on the base of $Q1$. The bias on $Q2$, in turn, is controlled by Zener current through $R2$.

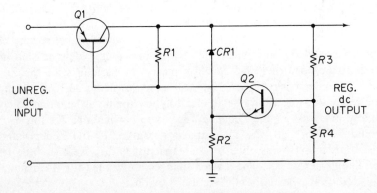

Figure 17-6. Electronic Series Regulator

An increase of input voltage, or an increase in load resistance, tends to cause an increase in output voltage. Either of these actions causes an increase in Zener current. The larger Zener current increases the voltage drop across $R2$, causing the emitter of $Q2$ to become more positive. This action reduces the forward bias on $Q2$, causing it to conduct less. The smaller collector current from $Q2$ causes less current through $R1$ and allows the base of $Q1$ to go more positive. Since $Q1$ is a *PNP* transistor, making the base more positive reduces its forward bias. The reduced bias causes an increase in the internal resistance of $Q1$ and a larger voltage drop across it. This increased series voltage drop offsets the original increase, and the output voltage holds at a steady level.

A decrease in input voltage, or a decrease in load resistance, tends to cause a decrease in output voltage. Either of these actions causes a decrease in Zener current. A smaller Zener current through $R2$ allows the emitter of $Q2$ to become less positive, and this increases its forward bias. The current through $Q2$ increases and causes more current through $R1$. The larger current through $R1$ causes the base of $Q1$ to become less positive. This is an increase in forward bias for $Q1$, and that causes its internal resistance to decrease. The smaller internal resistance results in a smaller series voltage drop across $Q1$. The decrease in series voltage drop offsets the original decrease, and the output voltage is held to a steady level.

The output is a fixed value of regulated dc within a specified tolerance.

### 17.4.2 Selecting Component Values

The series transistor, $Q1$, should be a heat-sink-mounted power transistor. It must be capable of handling the full load current with the maximum input voltage on the collector. Transistor $Q2$ needs to be capable of handling about half the maximum load current, although it will never have to do so under normal conditions. For example, if the maximum input voltage is 30 V and the maximum load current is 3 A, $Q1$ must be capable of handling 270 W and $Q2$ 135 W. Choose transistors with higher wattage ratings to allow an extra margin for safety.

The values of $R2$, $R3$, and $R4$ are selected for proper operation of $Q2$. The combined resistance of $R3$ and $R4$ is in parallel with the load resistance and must be very large in respect to the maximum load resistance. The current through $R3$ and $R4$ should never exceed 5 percent of the load current. Once a value has been established for $R2$, the trial value for $R1$ should be $0.1 \times R2$.

The Zener diode must have a breakdown voltage that is 1 or 2 V less than the desired regulated output voltage level.

### 17.4.3 Adjustable Output Regulator

The series regulator represented in Figure 17-6 can be designed for adjustable output capabilities by adding one more resistor. The added resistor should be a potentiometer, and it should be connected as illustrated in Figure 17-7. The potentiometer setting can be changed by a screwdriver adjustment, and this alters the fixed voltage level on the base of $Q2$. Moving the arm toward $R3$ increases the forward bias on $Q2$ and causes the regulator to produce a higher level of regulated output voltage. Moving the arm toward $R4$ decreases the forward bias on $Q2$ and reduces the level of regulated output.

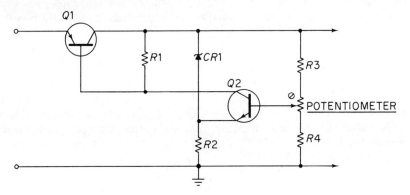

Figure 17-7.  Adjustable Output Regulator

## 17.5 Feedback Regulator

Transistors in a Darlington pair can be connected to form a *feedback regulator*. A single regulator of this type can compensate for *either* a change in input voltage or a change in output load resistance, or it can be made to handle *both* conditions. If the load current is fairly steady but the input voltage is subject to vary, the input regulator will suffice. If the load is subject to change but the input voltage is reasonably steady, the output regulator is better suited to the task.

### 17.5.1 Output Feedback Regulator

The schematic in Figure 17-8 illustrates the connection of a Darlington pair to regulate the output voltage under changing load conditions. Transistors $Q1$ and $Q2$ are connected as a Darlington pair. The remainder of the circuit is to control the bias on these two transistors.

## Voltage Regulators

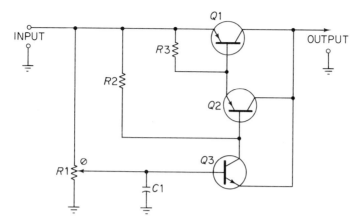

Figure 17-8. Output Regulator

Potentiometer $R3$ is set to the desired output level and the regulator holds the output to this level within a tolerance of $\pm 0.5$ V. The output voltage is connected to the collector of $Q1$ and $Q2$ and to the emitter of $Q3$.

An increase in load resistance results in a smaller load current which causes the emitter of $Q3$ to go more positive. More positive on the emitter decreases the forward bias and causes less collector current. Less current through $R2$ allows the base of $Q2$ to go more positive. Since $Q2$ is a *PNP*, the increased positive on its base reduces the emitter current. Less current through $R3$ allows the base of $Q1$ to go more positive, reducing its forward bias and increasing its internal resistance. The voltage drop across $Q1$ increases and offsets the original increase. This holds the output to a steady level.

A decrease in output load resistance results in a larger load current. This increases the forward bias on all transistors. The internal resistance of $Q1$ is reduced and less voltage is dropped across it. This reduced voltage across $Q1$ offsets the original decrease, and the output is held to a steady level.

### 17.5.2 Selecting Component Values

Transistors $Q1$ and $Q2$ may be obtained in a Darlington pair compound, or they can be two separate, well-matched transistors. All three transistors must be capable of handling the total input voltage. The series transistor, $Q1$, must be capable of handling total load current and a power equal to input voltage times load current. The power requirements for $Q2$ and $Q3$ are more modest. The power rating for $Q2$ should be about half that of $Q1$, and $Q3$ about 0.2 of $Q1$.

The value of $R1$ must be large enough to limit the current through it to less than 5 percent of the total load current. Trial values for $R2$ and $R3$ can be estimated in respect to the value of $R1$, as follows:

$$R2 = 0.1 \times R1$$
$$R3 = 0.01 \times R1$$

The center arm of $R1$ is now set to the desired regulated output level. If the output differs from this setting by more than 0.5 V, alter the value of $R2$ and $R3$ until the output is within tolerance.

Capacitor $C1$ is a filter capacitor. Its purpose is to prevent transients of short duration from affecting the potential on the base of $Q3$. Its value should be at least 500 $\mu$F. The voltage rating of $C1$ should be at least 1.3 times the input voltage.

### 17.5.3  Input and Output Regulator

The feedback regulator represented in Figure 17-8 does not compensate for changes in input voltage. However, the circuit can easily be modified to handle changes in both input voltage and output load resistance. The modification is accomplished by adding $R4$ and a Zener diode as illustrated in Figure 17-9.

The Zener diode in this regulator is the control on the output voltage level. The breakdown voltage for this diode should be only 1 or 2 V less than the regulated output level. The value of $R1$, as previously discussed, should limit the current through it to less than 5 percent of

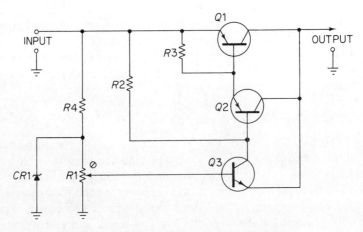

Figure 17-9.  Input and Output Regulator

the output current. But now the voltage across $R1$ is different. The Zener holds the voltage across $R1$ to the Zener voltage.

The current through $R1$ and the Zener current go through $R4$. The value of $R4$ is based on the current through $R1$ and the minimum voltage across $R4$. Figure the Zener current as about one third of the current through $R1$, and the value of $R4$ becomes

$$R4 = \frac{V_{in(min)} - V_{Zener}}{1.3 \times I_{R1}}$$

For example, when minimum input voltage is 20 V, Zener voltage is 15 V, and current through $R1$ is 10 mA, the resistance of $R4$ should be

$$R4 = \frac{5 \text{ V}}{13 \text{ mA}}$$
$$= 385 \ \Omega$$

Since all changes affect the Zener current, the maximum Zener current and power rating must be based on the maximum input voltage. The maximum Zener current is actually the fixed current through $R1$ subtracted from the maximum current through $R4$. Choose a Zener that can handle this current plus a margin for safety.

## 17.6 Dual-Output Regulator

Once a regulator has been designed, it is a simple matter to provide for more than one regulated output. This practice represents a maximum of economy and circuit utilization in areas requiring different levels of regulated voltage.

### 17.6.1 Sample Circuit

Figure 17-10 is a sample circuit for a series regulator with two levels of regulated output voltage. The standard regulator consists of $Q1$, $Q2$, $CR1$, $R1$, $C1$, and $R2$. With a 12-V to 15-V input this regulator will produce a 10-V output within a very close tolerance despite reasonable changes in both input voltage and output load resistance. How to select values of the parts for this regulator has been adequately discussed; so the trial values are simply listed as follows:

$$R1 = 2200 \ \Omega, 0.5 \text{ W}$$
$$R2 = 220 \ \Omega, 0.5 \text{ W}$$

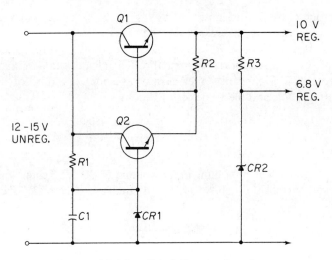

Figure 17-10.  Dual-Output Regulator

$C1 = 50\ \mu\text{F}$, 12 V, electrolytic

$CR1 = 9\ \text{V}, 0.1\ \text{A}$

### 17.6.2  Added Section

The added section, consisting of $CR2$ and $R3$, is actually a Zener shunt regulator to reduce the regulated output from 10 V to 6.8 V. Zener $CR2$ must have a breakdown voltage of 6.8 V with a 0.25-W power rating. So its maximum current rating should be about 30 mA.

The value of $R3$ should be such that the current through it is limited to load current plus a safe value of Zener current. The load current for this 6.8-V output plus the Zener current must drop exactly 3.2 V across $R3$. If the average load current is 12 mA, a flexible Zener current is about 6 mA, for a total current of 18 mA. So a reasonable value for $R3$ is 180 Ω, with a 0.5 W power rating. This allows plenty of room for load current changes without losing regulation.

## 17.7  Voltage Regulator for Heavy Current

In some cases, a series voltage regulator feeds loads that require more current than one series transistor can safely supply. The current capacity of a regulator can be extended by connecting two or more series dropping transistors in parallel with each other.

### 17.7.1 Sample Circuit

Figure 17-11 is a schematic of a sample voltage regulator that is designed to handle heavy currents. The series dropping transistors are Q3, Q4, and Q5. The forward and reverse bias for the three transistors are identical. Changing the bias on one changes the bias on all three. The internal resistance varies inversely with the forward bias.

The Zener diode, CR1, provides a fixed reference voltage for a wide range of load current. This diode works with Q1, Q2, Q6, and Q7 to control the bias on Q3, Q4, and Q5.

An increase in the input voltage, or an increase in output load resistance, tends to cause an increase in output voltage level. The Zener current increases, causing the base of Q6 to go more positive. The collector current of Q6 increases, causing the base of Q1 to go less positive. Transistors Q1 and Q2 conduct less, and this reduces the forward bias on Q3, Q4, and Q5. The resistance of Q3, Q4, and Q5 increases and more voltage drops across them. This larger series voltage drop offsets the original increase, and the output is held to a steady level.

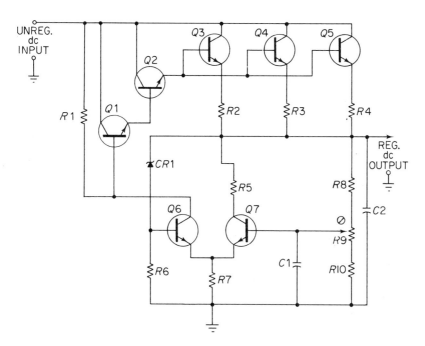

Figure 17-11. Voltage Regulator for Heavy Current

A decrease in the input voltage, or a decrease in output load resistance, tends to cause a decrease in the output voltage level. This causes a decrease of Zener current, and the base of $Q6$ goes less positive. The collector current of $Q6$ becomes smaller and causes the base of $Q1$ to go more positive. Transistors $Q1$ and $Q2$ conduct more and increase the forward bias on $Q3$, $Q4$, and $Q5$. The internal resistance of this combination decreases and less voltage is dropped across them. The smaller series voltage offsets the original decrease and holds the output voltage to a steady level.

Resistor $R9$ is a potentiometer for adjusting the level of regulated voltage output. Moving the arm of this resistor alters the bias on $Q7$ and $Q6$ without disturbing the Zener circuit. Moving $R9$ to a more positive level increases the forward bias on $Q7$ and decreases the forward bias on $Q6$. Transistor $Q6$ conducts less, $Q1$ and $Q2$ conduct more, and the resistance of $Q3$, $Q4$, and $Q5$ decreases. This action allows the output to be regulated at a higher level. Moving the arm of $R9$ to a less positive position will reverse these actions and cause the output to be regulated at a lower level.

### 17.7.2 Selecting Component Values

Transistors $Q3$, $Q4$, and $Q5$ should be power transistors mounted with heat sinks. The Zener should have a breakdown voltage equal to about half the desired regulated output. Resistors $R2$, $R3$, and $R4$ must have a very small value but a relatively high wattage rating; 0.1 $\Omega$ each is about right and never more than 0.5 $\Omega$.

The following list will serve as trial components under the conditions indicated:

Input voltage: 40 to 50 V.
Output voltage: adjustable from 22 to 30 V.
Load current: variable from 0 to 10 A.
$C1 = 1\ \mu F$, 25 V, paper.
$C2 = 100\ \mu F$, 50 V, electrolytic.
$CR1 = 12$ V, 80 mA.
$R1 = 1200\ \Omega$, 0.5 W.
$R2$, $R3$, and $R4 = 0.1\ \Omega$, 0.5 W each.
$R5 = 270\ \Omega$, 0.5 W.
$R6 = 2000\ \Omega$, 0.5 W.
$R7 = 570\ \Omega$, 0.5 W.
$R8$, $R9$, and $R10 = 1000\ \Omega$, 0.5 W each.

SECTION SEVEN

# TESTING AND TROUBLESHOOTING

The designer who constructs his own circuits must expect to spend a large portion of his time testing and troubleshooting his creations. This section includes hook-up and testing with commonly available test equipment that should facilitate the testing process. The troubleshooting procedures included here are also offered as a time-saving device. It is hoped that these tips will allow the designer to solve his problems faster and have more time for creating new circuits.

CHAPTER EIGHTEEN

# Bench Testing

No attempt has been made to cover all types of circuits nor all possible situations. Circuits and conditions have been selected which offer wide application as well as being of a type that the designer commonly encounters.

## 18.1   Measuring Circuit Characteristics

This section includes test circuits for measuring such characteristics as inductance, capacitance, impedance, and resonance. Circuits have been selected randomly with no attempt to concentrate on a specific type. However, the tests and measurements have a general application that makes them adaptable to many types of circuits.

### 18.1.1   Measuring Parallel Resonance

This test is a dynamic measurement of the resonant frequency of a parallel tank circuit. It is more realistic if the tank is measured in its own circuit. With the signal generator furnishing the signal, the test may be very simple, as indicated in Figure 18-1. The circuit illustrated is a

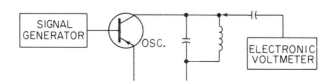

Figure 18-1.   Measuring Parallel Resonance

low-frequency circuit as evidenced by the coupling capacitor in the voltmeter lead. The circuit is easily adapted to intermediate and radio frequencies by using an appropriate probe for the meter input.

The procedure here is very basic. Isolate the oscillator or amplifier to be tested from its neighboring circuits, but leave the normal operating potentials present. Connect the signal-generator output into the input circuit of the test stage. Connect the electronic voltmeter to the tank circuit being tested.

Start the signal generator on a low frequency, well below the suspected resonant frequency of the tank circuit. Gradually increase the frequency while watching the output voltage on the voltmeter. As the resonant frequency is approached, the voltage will start to rise. This is caused by oscillations in the tank circuit.

There should be a fairly sharp peak when the resonant frequency of the tank is reached. As frequency increases above resonance, the voltage will drop. Tune the input frequency until the voltage is at its maximum point. The frequency indicated on the dial of the signal generator is the resonant frequency of the tank circuit.

Sometimes it is possible to measure frequencies above the upper limits of the signal-generator frequency. This is accomplished by using harmonics of the basic generator frequency to activate the test circuit. This method is not always reliable because of the low power in the harmonic frequencies. The power of the resonant harmonic may be too low to produce distinctive changes in the voltage reading.

### 18.1.2 Measuring Inductance and Capacitance

If either inductance or capacitance of a tank circuit is known, the other may be determined by carrying the resonance test one step further. Using the resonant frequency and the value of the known component, the other can be calculated by

$$f = \frac{1}{2\pi\sqrt{LC}}$$

For example, when a 2-pF capacitor produces a resonant frequency of 5 MHz, the inductance of the coil is

$$L = \frac{(0.159/f)^2}{C}$$
$$= \frac{(0.159/5 \times 10^6)^2}{2 \times 10^{-12}}$$
$$= \frac{(0.032 \times 10^{-6})^2}{2 \times 10^{-12}}$$

# Bench Testing

$$= 5 \times 10^{-4}$$
$$= 0.5 \text{ mH}$$

When the inductance is the known quantity, the same formula can be used to calculate the capacitance. It is only necessary to exchange places with the $L$ and $C$:

$$C = \frac{(0.159/f)^2}{L}$$

Since the quantities are generally frequency in MHz, inductance in $\mu$H, and capacitance in pF, considerable work may be saved by keeping these units in mind during calculation. The formulas would then become

$$L = \frac{25,281}{C \times f^2}$$

and

$$C = \frac{25,281}{L \times f^2}$$

where $L$ is the inductance, in $\mu$H; $C$ is the capacitance, in pF; and $f$ is the frequency, in MHz.

### 18.1.3 Measuring Distributive Capacitance

Every circuit has a certain amount of distributive capacitance. This becomes especially bothersome in transformer and inductive coupling circuits. In fact, every coil has enough distributive capacitance to become a parallel-resonant tank circuit at some frequency. This fact enables measurement of the distributive capacitance.

The test circuit is similar to that in Figure 18-1, but there is no physical capacitor across the coil. Resonant frequency is obtained in the manner previously described. The resonant frequency and the inductance of the coil are then used to calculate the distributive capacitance.

### 18.1.4 Measuring Series Resonance

The same equipment used for parallel-resonant circuits can be used for measuring the resonant frequency of a series circuit. The test circuit is similar to that shown in Figure 18-2. The signal generator and electronic voltmeter are both connected parallel to the resonant circuit. The

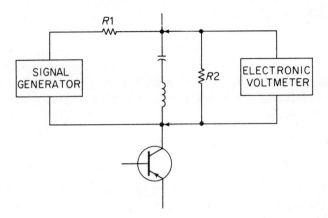

Figure 18-2.  Measuring Series Resonance

resistor ($R_1$) in series with the signal generator is to limit the current when resonance is reached. The value of this resistor is not critical, but it is necessary to prevent generator overload as the impedance of the *LC* circuit approaches zero.

The parallel resistor ($R_2$) is necessary to load the circuit and broaden the resonant response. Without this resistor the peak may be too sharp to detect on the meter. The value is not critical, but the smaller the resistor, the sharper the response. A 50-k$\Omega$ resistor will probably be about right.

Start the signal generator at a low frequency and gradually increase the frequency while watching for a decrease in the voltage indication. The voltage will start to drop as the resonant frequency is approached. Tune the generator for a minimum voltage indication. The output frequency of the signal generator is now the same as the resonant frequency of the *LC* circuit.

### 18.1.5  Avoiding False Indications

In any resonance measurement, it is possible to obtain resonance indications on harmonics of the resonant frequency. This is one reason for starting the generator at a frequency below resonance. It is also possible to obtain resonance indications as the generator is tuned through a submultiple of the resonant frequency. These indications are less likely and less pronounced than the harmonics, but they can be deceptive.

Both harmonic and submultiple indications can be avoided when the approximate resonant frequency is known. The signal generator is then set just below the suspected point of resonance. In many cases, the approximate frequency is not known. Then the only procedure is to start the frequency low and beware of small indications. After locating what is apparently the correct resonant frequency, it is a good idea to continue raising the frequency until another indication is obtained. If this is a harmonic indication, it will be much less pronounced than the first one. If all readings are questioned in this fashion, there is little chance of being taken in by a false indication.

### 18.1.6 Measuring the Bandwidth

Once resonant frequency has been established, it is little additional effort to determine the bandwidth. However, when this is the intent, the input and output connections become more critical. Since loading broadens the resonance-response curve, the loading must be minimized to enable accurate determination of the bandwidth. One method of accomplishing this is indicated in Figure 18-3.

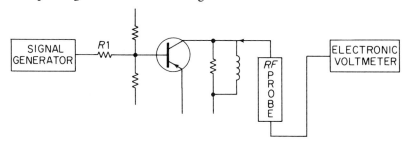

Figure 18-3. Measuring the Bandwidth

The output of the signal generator is coupled through an isolation resistor ($R_1$) into the test circuit. This resistor minimizes the loading effect of the signal generator, but the resistor value is not critical. The electronic voltmeter has a minimum loading effect anyway, and in many cases the meter input will be coupled through an RF probe.

The signal generator is now tuned to the resonant frequency in the manner previously described. The bandwidth extends between the low half-power point and the high half-power point. These points are determined by locating a frequency that provides a voltage reading that is equal to 0.707 × the maximum. The voltage is maximum at the resonant frequency ($f_r$). Note this reading and calculate 0.707 of its value. Raise the frequency of the signal generator until the voltage indication drops to the calculated 0.707 point. This is the high half-power point. Note this frequency and designate it as $f_1$.

Reduce the frequency of the signal generator back to resonance, and beyond, until the 0.707 value is encountered again. This is the low half-power point. Note this frequency and designate it as $f_2$. The relationship of these three frequencies is illustrated in Figure 18-4. The bandwidth is calculated by subtracting $f_2$ from $f_1$. For example, when high half-power frequency is 460 kHz, and low half-power frequency is 450 kHz, the bandwidth is

$$BW = f_1 - f_2$$
$$= 460 - 450 = 10 \text{ kHz}$$

The terms *bandwidth* and *bandpass* are often confused. The previous explanation should have clarified the meaning of bandwidth. Bandpass is simply the actual frequencies contained within the bandwidth. In the previous example, the bandpass is 450 to 460 kHz.

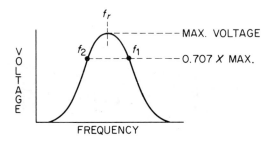

Figure 18-4. Frequency Response

### 18.1.7 Determining the Q of a Circuit

The $Q$ of a circuit is directly related to resonant frequency and bandwidth in the following manner:

$$Q = \frac{f_r}{BW} \qquad BW = \frac{f_r}{Q} \qquad f_r = BW \times Q$$

For example, a circuit with a resonant frequency of 455 kHz and a bandwidth of 10 kHz has a $Q$ as shown here:

$$Q = \frac{f_r}{BW}$$
$$= \frac{455 \text{ kHz}}{10 \text{ kHz}}$$
$$= 45.5$$

## 18.2 Measuring Switching Characteristics

Many transistors, both bipolar and field-effect, are designed for *switching* applications. It is important to match the characteristics of the switching component to the circuit where it will be used. The switching characteristics are listed on the data sheet, and they can be measured by the use of an oscilloscope and a square-wave signal generator. The oscilloscope should be a dual trace with good transient characteristics and a wide-frequency response. The horizontal deflection on the oscilloscope should be time-calibrated.

### 18.2.1 Measuring Bipolar Transistors

In order to measure the switching characteristics of a transistor, it is necessary to place it in a properly biased circuit which closely simulates operating conditions. The signal generator then provides input pulses, and the input and output are compared on the oscilloscope. A sample test circuit is illustrated in Figure 18-5.

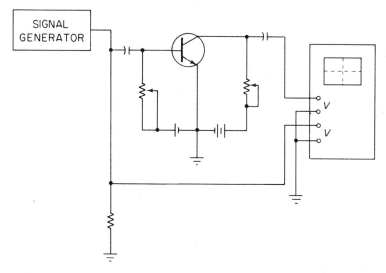

Figure 18-5. Measuring Bipolar Switching Characteristics

The transistor is properly biased and the voltages are adjusted to the values specified in the data sheet. A pulse from the signal generator is applied directly to one vertical channel of the oscilloscope. The same pulse is applied to the other vertical channel after being processed through the transistor. The amplitude and pulse width of the signal-generator output should be set to the values shown on the data sheet.

The signal-generator output pulse appears on one trace, and the transistor output appears on the other trace. The critical times such as rise time, delay time, storage time, and fall time can now be measured along the horizontal axis. These times must be compared to times indicated on the data sheets to determine whether or not they are within tolerance.

### 18.2.2 Measuring FET Switching Characteristics

The switching characteristics of the field-effect transistor can be measured by practically the same procedure just described for standard transistors. The circuit for the FET is slightly different, but the signal generator and oscilloscope are connected in the same fashion. This is illustrated in Figure 18-6.

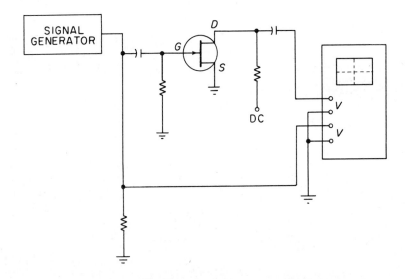

Figure 18-6. Measuring FET Switching Characteristics

The same times are critical for FETs as for any switching transistor (rise time, delay time, storage time, and fall time). These times are specified on the data sheets, and measurements are intended to verify these data. These measurements are taken as previously described, and the results are compared to the specified data.

## 18.3 Audio Circuits

The signal generator is an indispensable piece of equipment when designing, testing, or troubleshooting audio circuits. A great many tests and measurements may be made to determine the exact condition of a stage. Several simple measurements are described in this section. The text and drawings are audio-oriented, but with slight variations the same general procedures and equipment may be used for measurements in intermediate and radio-frequency circuits. Where there is a choice of instruments, the simplest one should be used; the same is true of the procedures.

### 18.3.1 Measuring the Frequency Response

The *frequency-response* measurement is essential for filters, coupling circuits, amplifier stages, and overall audio sections. An amplifier stage is used as an example. A test circuit for this stage is illustrated in Figure 18-7. The purpose of this measurement is to determine the band of frequencies that this stage can efficiently handle. Since it is an audio stage, it is expected to have a flat response from about 50 Hz to something in excess of 20 kHz. A well-designed stage may have a relatively flat response from 15 Hz to more than 50 kHz. The signal generator is coupled to the input of the stage, and the electronic voltmeter is connected across the output load resistor. For the best results, the amplifier should be in its own operational circuit. If this is not practical, the test circuit should duplicate the operational circuit as nearly as possible.

The signal generator is set to a low-frequency sine-wave output, and the amplitude is set to a value representative of a reasonable audio signal. This amplitude is held constant as the frequency is slowly increased.

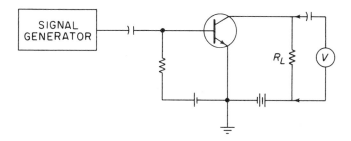

Figure 18-7. Measuring the AF Response

The voltmeter is set on ac and has no indication until the signal feeds through the amplifier. As the sine-wave frequency approaches the audio band, a small voltage indication will appear on the voltmeter. The voltage will level off to a steady reading as the frequency reaches a value where the amplifier has maximum response. This should happen at about 20 Hz.

The constant-voltage reading will hold until the frequency approaches the top of the response band. As the frequency increases above this value, the voltage slowly drops off.

A plot of the voltage and frequency on a simple graph such as that in Figure 18-8 will provide a clear picture of the frequency response of this amplifier stage.

The actual response of the stage is the same as the bandpass, the frequencies being between the low and high half-power points. Since the measurements and graph are in volts, the 0.707 voltage points may be used. The graph in the illustration shows a frequency response from 12 Hz to 75 kHz, and the top of the curve is reasonably flat all the way. A response such as this is obtained from an exceptionally responsive stage.

### 18.3.2 Measuring Harmonic Distortion

A scientific analysis of harmonic distortion utilizes special equipment and may be a prolonged operation requiring a great deal of skill. Measuring the percentage of harmonic distortion, however, may be a simple process performed by standard shop instruments. One method is illustrated in Figure 18-9. The procedure is designed to measure the

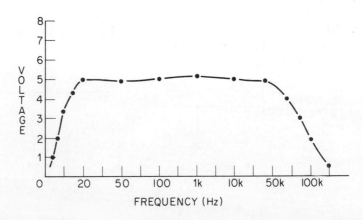

Figure 18-8.  Plotting the Frequency Response

amplifier output under two conditions: one with a filter and one without a filter. Without the filter, the output is the resultant of the fundamental frequency and all the harmonics. With the filter, the output is a result of the harmonics only. The filter eliminates the fundamental frequency.

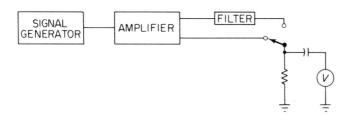

Figure 18-9. Measuring Harmonic Distortion

The signal generator is set for a sine-wave output with a reasonable voltage value. The filter and the signal generator are then tuned to the same frequency. Some filters can be adjusted and some cannot. But the frequency of the signal generator is adjustable. This may be accomplished by connecting the signal generator directly to the filter and tuning for minimum indication on the voltmeter.

Then with the circuit connected as shown in Figure 18-9, set the signal generator for a strong signal, but be careful not to overdrive the amplifier. Now take one voltage reading without the filter ($V_1$) and another reading with the filter ($V_2$). The harmonic distortion percentage is then calculated by

$$\text{harmonic distortion percentage} = \frac{V_2}{V_1} \times 100$$

For example, when the voltage without a filter is 150 mV, and with a filter it measures 5 mV, the percentage of harmonic distortion is

$$\text{harmonic distortion percentage} = \frac{V_2}{V_1} \times 100$$

$$= \frac{5 \text{ mV}}{150 \text{ mV}} \times 100$$

$$= 0.033 \times 100$$

$$= 3.3\%$$

### 18.3.3 Measuring Input Impedance

The signal generator and the electronic voltmeter may be used to measure input impedance by the resistance substitution method. This method is highly accurate when the impedance is mostly resistance and is an approximation when the impedance is mostly reactance. Since impedance must always be matched for maximum transfer of energy, an easy way to measure impedance is a great asset. A test circuit for measuring input impedance is illustrated in Figure 18-10.

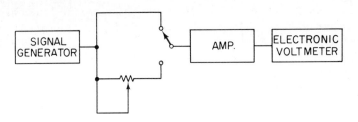

Figure 18-10.  Measuring Input Impedance

The signal generator is set to the type of signal and frequency desired. These should be, as nearly as possible, duplications of what the amplifier will be expected to handle in its own circuit. The test circuit is connected as shown, with the signal generator directly connected to the amplifier input and the electronic voltmeter connected across the amplifier output.

The amplitude of the signal-generator output is adjusted to obtain a strong reading on the voltmeter, at least a half-scale deflection. Record this reading. The switch is then changed to place the rheostat in series with the amplifier input. The rheostat is adjusted until the voltmeter indicates exactly half the previous value. Remove the rheostat and measure its resistance; this is equal to the input impedance of the amplifier.

The *rheostat* test circuit loses accuracy when the input impedance is very low (only a few hundred ohms). In this case, a *potentiometer* should be used as illustrated in Figure 18-11.

The series resistor ($R_1$) should be at least 10 times the value of the input impedance; otherwise, its value is not critical. The potentiometer ($R_2$) is set to maximum resistance, with points 1 and 2 shorted directly together. The signal-generator output level is adjusted for a convenient reading on the voltmeter, and this reading is recorded. Now $R_2$ is adjusted until the voltage indication drops to half the recorded level. The resistance of $R_2$, measured between points 2 and 3, is the same as the input impedance of the amplifier.

# Bench Testing

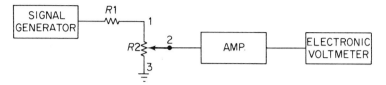

Figure 18-11. Measuring Low Input Impedance

### 18.3.4 Measuring Output Impedance

The output impedance of a circuit or a stage can be measured with a slight variation of the previous procedure. The instruments are connected as shown in Figure 18-12. The signal generator is connected to the input of the stage being tested, and the electronic voltmeter is connected across the output. The signal generator is set to the desired signal and frequency. The amplitude of the signal-generator output is then increased until a convenient value of ac voltage is indicated on the voltmeter. Record this voltage reading.

Now the rheostat is connected parallel to the amplifier output. The rheostat is adjusted until the voltage drops to half the recorded value. The rheostat is then removed from the circuit and measured. The resistance of the rheostat is equal to the output impedance of the stage under test.

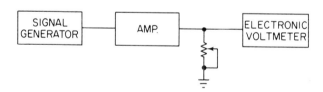

Figure 18-12. Measuring Output Impedance

## 18.4 Measuring Magnetic Components

There are certain checks and measurements that can be made on electromagnetic devices such as microphones, speakers, and transformers by using the signal generator. Microphones and speakers need to be checked for distortion and frequency response. Sometimes it is necessary to check transformers for phase relationship, polarity marking, and percentage of regulation. All of these tests, and many others, can be made with a signal generator, a voltmeter, and an oscilloscope.

### 18.4.1 Checking Microphones

This procedure assumes that a matching speaker of known quality is available. The frequency response of the *microphone* and the distortion of the signal caused by the microphone can both be checked with one simple arrangement. The equipment and instruments are arranged as indicated in Figure 18-13.

The signal generator is connected to a speaker that has been verified as having good frequency response and being free of distortion. The signal generator is set for a sine-wave output and adjusted for a desired amplitude. This output is connected to one vertical channel of a dual-trace oscilloscope as well as to the speaker.

The microphone under test is connected to an amplifier and placed in a position to enable good pickup of the sound from the speaker. The microphone should be shielded from all other sound as nearly as possible. The output of the amplifier is connected to the other vertical channel of the oscilloscope.

The two waveforms can be viewed simultaneously, and distortion is readily apparent. The signal generator can now be tuned through the rated frequency band of the microphone while watching the display for changes in amplitude or evidence of distortion. The amplitude of the signal generator output must, of course, be held constant while changing the frequency. Any change in amplifier output amplitude indicates that the microphone is responding differently to the different frequencies. A decrease in amplitude shows a low response to the frequency.

If there is any doubt that the amplifier is at fault instead of the microphone, a further check is indicated. The amplifier can be checked by removing both microphone and speaker and connecting the signal generator directly to the amplifier input.

### 18.4.2 Checking Speakers

The same procedure described for checking microphones is equally effective for checking speakers. In this case, the microphone and am-

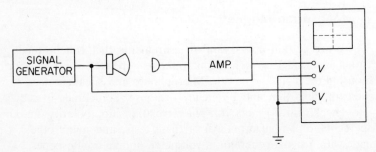

Figure 18-13. Checking a Microphone

plifier must be of a known reliable quality. The equipment and instruments are connected as shown in Figure 18-13. Now any distortion or lack of frequency response is caused by the speaker.

### 18.4.3 Determining Phase Relationships Between Windings of a Transformer

A transformer can be a puzzling device when it is disconnected from the circuit and the phase markings are not clear. Many perfectly good transformers have been discarded because the user was uncertain of the phase relation of primary to secondary. This need not happen. A signal generator and a voltmeter can be used to determine the phase relationship, and the check requires only a few moments.

The signal generator is connected to the primary winding and set to the desired frequency and amplitude. The voltmeter is used to measure the voltage from the secondary. The values of input and output voltages are recorded. Then connections are made as illustrated in Figure 18-14.

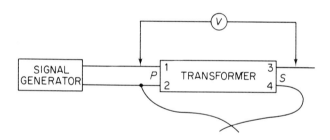

Figure 18-14.   Primary-to-Secondary Phase Check

Pins 2 and 4 are shorted together, and the voltage is measured between pins 1 and 3. If pins 1 and 3 are in phase, this reading will be equal to the difference between the first two (primary and secondary) measurements. If pins 1 and 4 are in phase, this reading will be the equal to the sum of the first two measurements. For example, with a primary voltage of 10 V and a secondary of 30 V, the following results are obtained.

1. If 1 and 3 are in phase, the voltage between pins 1 and 3 is 20 V.
2. If 1 and 4 are in phase, the voltage between pins 1 and 3 is 40 V.

### 18.4.4 Measuring Transformer Regulation

The *regulation* of a transformer is its ability to maintain a constant output voltage with a changing load. The percentage of regulation can be measured with a test arrangement similar to that in Figure 18-15.

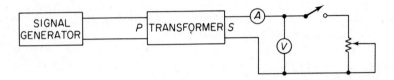

Figure 18-15. Measuring Regulation

One output voltage reading is taken with the secondary open. If this reading changes only slightly when the switch is closed, regulation is good. The rheostat is then decreased until the ammeter indicates the maximum rated current for the transformer. The voltage indication should remain fairly constant during this change.

If no load voltage is designated as $V_1$ and the voltage under maximum current conditions (load voltage) is designated as $V_2$, the percentage of regulation can be calculated as follows:

$$\text{regulation percentage} = \frac{V_1 - V_2}{V_1} \times 100$$

$$= \frac{50 \text{ V} - 48 \text{ V}}{50 \text{ V}} \times 100$$

$$= \frac{2}{50} \times 100$$

$$= 4\%$$

The lower the regulation percentage, the better the regulation.

### 18.5 Measuring Antennas

Several very practical antenna measurements, such as those for antenna resonance, impedance, and gain, can be made with little trouble by utilizing common basic test instruments. These measurements can be performed with a signal generator and an ordinary electronic multimeter.

### 18.5.1 Measuring the Resonant Frequency

The resonant frequency of an antenna is easily measured with a signal generator and an ammeter. The test circuit is illustrated in Figure 18-16. The signal generator is set for the desired output level and a frequency well below the suspected resonant frequency of the antenna. The amplitude of the signal-generator output is then held constant as the frequency is gradually increased. The frequency that causes the maximum current indication is the resonant frequency of the antenna.

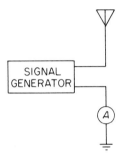

Figure 18-16.  Measuring the Resonant Frequency

### 18.5.2 Measuring Antenna Impedance

The impedance of an antenna, along with its associated transmission line, can be easily measured by the resistance substitution method. A test circuit for this measurement is illustrated in Figure 18-17.

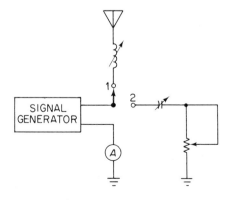

Figure 18-17.  Measuring Impedance

The signal generator is adjusted to the resonant frequency of the antenna as previously described, and the current indication is recorded. Now the switch is changed to position 2, and the capacitor is set for maximum current. The rheostat is now adjusted until the previously recorded indication is obtained on the ammeter. At this time, the resistance of the rheostat is equal to the impedance of the antenna and its associated transmission line. This is the radiation impedance.

An assumption was made here that the antenna was tuned with an inductor. Therefore, a tunable capacitor was used in the substitution circuit. If the antenna is tuned with a capacitor, the substitution circuit should contain a tunable inductor. The procedure is the same in either case.

### 18.5.3 Measuring Antenna Gain

The gain of a receiver antenna is a ratio between the voltage developed on the antenna and the voltage developed on a reference dipole of the same size. The reference dipole is constructed of a thin wire, and its gain provides the reference of 0 dB. The antenna gain is generally plotted on a graph in terms of decibels above this reference level. When the voltage on the antenna is designated as $V_1$ and that on the reference dipole as $V_2$, then

$$\text{gain} = 10 \log \frac{V_1}{V_2}$$

Figure 18-18 illustrates a test circuit for making this measurement. The signal generator is connected to a transmitter antenna and takes the place of the transmitter during this test. The signal generator is set to the desired frequency and amplitude.

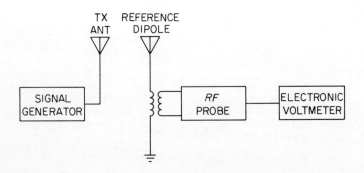

Figure 18-18. Measuring Gain

The reference dipole is then placed in a predetermined position, and the voltage induced on it is measured with an electronic voltmeter. The voltage is recorded as $V_2$.

The reference dipole is then removed, and the test antenna is put in the exact place and position previously held by the dipole. The induced voltage on the antenna is now measured and recorded as $V_1$. Care must be exercised to put the antenna and reference dipole in exactly the same place and position during the voltage measurements. It is also assumed that all other conditions are identical.

CHAPTER NINETEEN

# Troubleshooting Techniques

Troubleshooting is a logical procedure that should be applied to the isolation of trouble in all types of electronic equipment. There is little difference in the procedure between troubleshooting for a single stage, a transistor radio, or for a complex electronic system. The technician who approaches all troubleshooting problems in a logical, systematic manner will consistently locate troubles faster than one who tries haphazard shortcuts. Occasionally a hit-or-miss method of troubleshooting will locate a trouble. A lucky guess may pay off once, but the chances are against it.

No amount of logic can replace a knowledge of the equipment, and no troubleshooting should be attempted without this knowledge. Armed with this knowledge and a logical approach, the technician can functionally divide any electronic equipment and test it in an orderly and professional manner.

## 19.1  Six-Step Procedure

The six-step procedure in troubleshooting is a standardized approach designed to save time and needless repairs. A great many highly proficient technicians and engineers probably cannot name these six steps, but it is a safe bet that they do apply them either consciously or unconsciously. Here are the six steps:

1. Recognizing symptoms.
2. Enhancing the symptoms.
3. Listing probable faulty functions.
4. Isolating the faulty function.

5. Localizing trouble to a specific circuit.
6. Isolating the faulty component.

### 19.1.1  Step 1—Recognizing Symptoms

The first step is the recognition of symptoms. A trouble symptom is a sign that the equipment has a disorder or a malfunction. This implies that there is a standard set of normal indications for comparison. The normal symptoms are usually pretty well impressed on the troubleshooter's mind from day-to-day observation. Slight deviations from the norm are probably recorded in areas where a record is maintained. Without knowledge of the equipment and its normal indications, any trouble symptoms short of a complete breakdown are likely to pass unnoticed.

The normal indications for a black-and-white television receiver are as follows:

1. A clear, properly contrasted picture of an actual scene.
2. The picture is centered within the vertical and horizontal boundaries of the screen.
3. A clearly understandable reproduction of voice and music sounds.

Possible abnormal indications or trouble symptoms of the same television receiver are:

1. Picture rolling vertically.
2. Lack of contrast.
3. Picture tearing.
4. Garbled or distorted sounds.

The symptoms just listed are readily *recognizable* because they are deviations from the normal indications. Symptoms in other equipment may be a trouble light, a low current reading, voltage out of tolerance, or a loss of power.

During the process of performing its assigned job, electronic equipment yields information that can be seen or heard. The experienced technician knows the normal sights and sounds. Thus, the senses of sight and hearing are the first to receive a trouble symptom. Trouble symptoms must be recognized as indications of abnormal performance.

Complete breakdown, of course, is the easiest trouble symptom to recognize. This is probably the one seen most frequently in the repair shop. The equipment simply does not function. Some of these failures are evidenced by: no sound from a speaker, no picture on a CRT, and zero indications on panel meters.

All audible and visual indications may be present, and yet the set may have abnormal performance. When the audible and visual indications do not conform to the manufacturer's specifications, the performance is *degraded*. Degraded performance is a forerunner of equipment breakdown and should be corrected at once. These degraded performance indications range all the way from near-perfect performance to near-total lack of performance. The equipment is still performing its job, but not as well as it should.

Whatever the nature of the trouble symptoms, they must be recognized as a first step in troubleshooting.

### 19.1.2 Step 2—Enhancing Symptoms

The second step is intended to extract additional information from the symptoms observed. A particular symptom may be caused by any one of several equipment faults. Jumping from here into schematic analysis can lead to involvement in unrealistic and unnecessary pursuits. The symptom needs to be aggravated and analyzed carefully to extract all the information that it is capable of giving. This is a process of symptom elaboration to obtain a more detailed description of the symptom.

Manipulation of operating controls and switches will produce a variety of changes in the circuit conditions. These changes will alter current or voltage readings by changing resistance or impedance of the circuit. This can cause detrimental effects in equipment performance. Observation of all indicators while making changes will, in many cases, reveal some definite details about the symptoms already observed.

During control manipulation, care must be exercised not to overcontrol. Faulty placement of controls may have caused the present trouble. Additional damage may result if controls are handled carelessly. Electronic circuit components have definite maximum values of voltage and current. If these values are exceeded, burned-out components could be the result. A knowledge of the circuit changes that take place when a control is turned will enable the troubleshooter to think ahead and avoid any damage the control could cause.

In many cases, the trouble symptom will be produced solely by an improper setting of a control or switch. This is not always a careless misadjustment; controls can be moved accidentally. When this happens, a proper adjustment of the controls corrects the trouble and results in a disappearance of the trouble symptom. Many television and radio receivers have been turned in for repair because the power cord was disconnected or some control was set incorrectly.

Even if there is no fault with the controls, the time spent checking them is time well spent. The controls affect a good many circuits and should furnish information for further defining the symptoms. For example, a stereo AM-FM receiver has distorted sound. When the control is moved from AM to FM, the symptom disappears. This eliminates all the FM circuits as possible causes of the trouble, and so hastens the solution of the problem.

Next, the observed symptoms must be fully evaluated in order to completely clarify the symptoms. The indications must be evaluated in relation to one another and in relation to the overall operation of the equipment. It is a good idea to record each bit of data as it is obtained. List the original symptoms; make notes on how the symptoms change as the controls are moved; if moving a control has no effect, make a note of that, too. Now sit back for a moment and digest the information before starting the next step.

### 19.1.3 Step 3—Listing Probable Faulty Functions

Steps 3 and 4 are not always needed; their performance depends on the complexity of the equipment. For instance, a small transistor radio receiver has only one function, so listing the probable faulty functions and isolating the faulty function are meaningless. In this case the troubleshooter should proceed directly to step 5 and start to localize the trouble to a specific circuit.

On the other hand, a transceiver could be considered as having at least six functions: sound pickup, modulation, transmitter, antenna assembly, receiver, and power supply. Any equipment that performs more than one function should be given the benefit of all six troubleshooting steps.

Therefore, we turn again to step 3. The symptoms have been recognized and enhanced by careful analysis. Now, what functions are disabled? Trouble in several places at the same time is not very likely. Is there a single functional area that could cause all the observed symptoms? The symptoms should be studied again—this time along with a functional block diagram of the equipment. Eliminate each functional area if it alone cannot account for all the observed indications, and those that remain comprise a list of suspects. This elimination process has reduced the problem to the realm of possibility. There remain a lot of circuits to be checked, but not the entire system.

Consider the diagram in Figure 19-1. This is a functional block diagram of a standard AM transceiver. Each block performs a specific function, so this is a symbolical representation of the functional units within the equipment. From this diagram and a few carefully analyzed symptoms, two, three, four, or five functions can be eliminated. For ex-

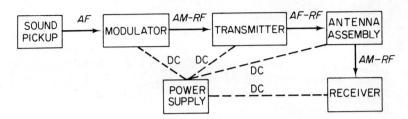

Figure 19-1. Transceiver Functional Diagram

ample, the original symptoms are the lack of reception and the fact that manipulation of the receiver controls has no effect. The power-on light and the dial lights of the receiver are still lit. Out of the six functional units, only three could be at fault: power supply, antenna, and receiver. That is a good start, but a further reduction of possible faulty functions may be possible.

The three suspect units should now be subjected to a test on a faulty unit selector. This selector is constructed as illustrated in Figure 19-2. Take each of the suspected units in turn and ask: "Can a fault in this unit cause the original symptom?" Knowledge of the equipment and the functional diagram should furnish enough information to answer this question with a definite "yes" or "no." If not, refer to schematics, wiring diagrams, and any other source of information. The answer must be correct. If the answer is no, eliminate that unit from further consideration. If the answer is yes, pose the second question: "Can a fault in this unit produce all the associated information?" Again the question is answered yes or no, after carefully considering all the facts. The answer must be accurate. If it is no, eliminate the unit from further consideration. If it is yes, keep the unit on the list of suspects.

Since the receiver controls had no effect on the symptom, the second question eliminates the antenna as a possible cause. A faulty antenna could block reception, but it could not prevent the volume control from varying the noise level. This step reduces the possible faulty units to the receiver and the power supply. These units get a positive yes as an answer to both questions; they remain as suspects.

The trouble is still unknown, but the total area for testing has been reduced by two-thirds. This will undoubtedly save a lot of time that would otherwise have been wasted in testing in areas that could not possibly be at fault. Each of the two remaining units could be at fault, but so far they are only suspect. Each unit has been selected by the analysis of technically valid evidence, but each is only a probable source of trouble. The next step in the procedure will narrow the search area still further.

# Troubleshooting Techniques

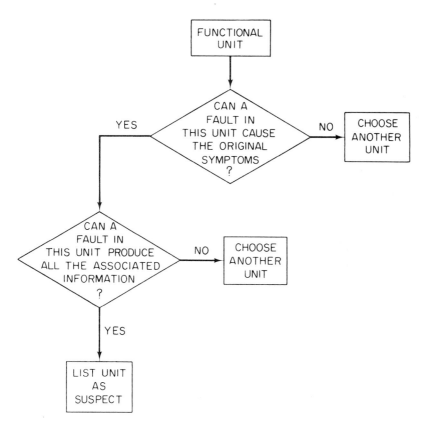

Figure 19-2. Faulty Unit Selector

### 19.1.4 Step 4—Isolating the Faulty Function

Step 4 is intended to save additional time by logically isolating the functional unit that is most probably faulty. Thus far the procedure has dealt with apparent and obscure performance deficiencies. This led to a logical selection of probable faulty units. No test instruments have been used, except built-in meters and indicators, and no dust covers or drawers have been removed. All actions have been either superficial or rational, but the scope of the problem has been drastically reduced. Now, from the list of probables, it is necessary to isolate the functional unit that is responsible for the failure.

Testing will be done during this step, but it should be of a limited nature. These tests, combined with the same logical process, should lead to a definite identification of the faulty function. Which unit on the list of suspects should be checked first? Consider the block diagram, accessibility of test points, and past experience with troubles of this nature. Consider which unit will yield the most information, no matter whether it proves good or bad. Remember that test points which require disassembly of the equipment for access should be avoided as a first choice. Further, considering past history, check if the past three failures of this type were caused by power-supply problems, and if so, check the power supply.

The results of each test should lead to the next logical test, and the combined tests should isolate the faulty function. This system will succeed if all actions up to this point have been accurate, but mistakes do occur at times. When the suspected faulty units are proved to be good, this shows a previous wrong conclusion some place along the line. Backtrack all the way to step 1. Reevaluate the data; the written notes will help. Locate the point where reasoning was faulty and pick up a new track. This type of action will be necessary only occasionally, provided that the first three steps have been followed carefully. When it is necessary, troubleshooters should be armed with extra data which enable them to zero in on the faulty area in a very short time. At least he now knows some units that cannot be at fault.

Even when preliminary tests indicate that a particular unit is bad, hasty conclusions should be avoided. Troubleshooters should draw on their knowledge of equipment and think back over the ground that they have covered. They should ask: "What trouble symptoms can be produced when this unit fails?" If this leads them back to the original symptoms and accounts for all the associated information, they can be sure that they have the right unit. The matching of the theoretical symptoms with the actual symptoms should leave no room for doubt.

### 19.1.5 Step 5—Localizing Trouble to a Specific Circuit

In the case of very simple equipment, troubleshooters may arrive at step 5 directly from step 2. With more complex equipment, they may arrive only after careful consideration of all aspects of the information uncovered through the first four steps. Now they can break out the test instruments and remove the dust covers. But when doing so, they know that they are not performing a useless task. They know that the trouble must exist in this functional unit.

If the faulty unit is a radio receiver, the test instruments should consist of a multimeter and an oscilloscope. Some other specialized instruments may be indicated for other units. But troubleshooters should not begin arbitrary measurements. They should first consult the *servicing block diagrams* in maintenance manuals. A sample of such a diagram is illustrated in Figure 19-3.

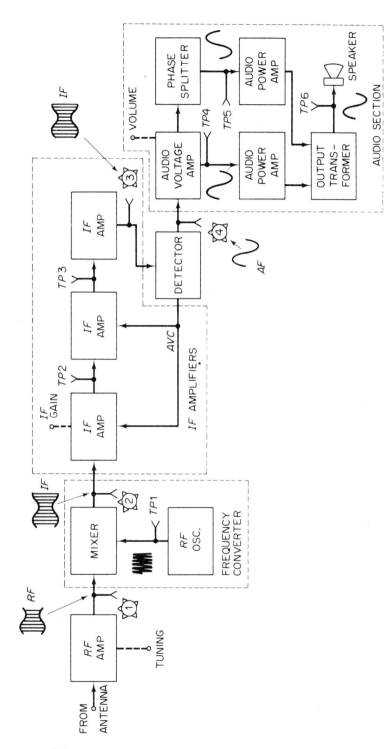

Figure 19-3. Servicing Diagram for a Receiver

The servicing block diagram separates the units into circuit groups. Each of these circuit groups performs an electronic subfunction that is vital to the task assigned to this functional unit. This diagram is for the receiver section of a transceiver, but there are other diagrams for all other functions. Notice that this diagram is broken into five groups: RF section, frequency converter section, IF section, detector section, and audio section. The task to be performed in troubleshooting in step 5 is to localize the fault to a circuit, but first it must be determined which of these five circuit groups contains the malfunction. This can be determined with no more than four readings on the oscilloscope.

Notice that there are four test points marked with stars. Only these points need be considered while narrowing the problem down to a circuit group. Should the readings be taken in numerical order? Not necessarily. Which measurements will eliminate the greatest number of circuits? Number 4 looks like a good bet.

Connect the oscilloscope to the test point, Star 4, and take a reading. If the audio signal is present, the job is finished. The bad circuit has to be in the audio circuit group. If the signal is not present, the audio group has been eliminated as a cause; it must be a good circuit. Next try Star 2. If the IF signal is present, the trouble is isolated to either the IF section or the detector. Now a check at Star 3 will pin it down. A signal here isolates the trouble to the detector. No signal here isolates the trouble to the IF section. Backing up for a moment, if *no signal* appears at Star 2, it eliminates all circuits from that point forward; the trouble has to be in either the RF section or the frequency-converter section. One more check will pin it down; take a reading at Star 1. A signal here labels the frequency-converter section as the bad circuit group. No signal at this point fixes the blame on the RF section.

The method used here in isolating the faulty circuit group is sometimes referred to as a *bracketing process*. At first the brackets enclosed the entire receiver. The brackets were then closed in after each measurement until they enclosed only the faulty circuit group. This method is further illustrated in Figure 19-4.

Group 1, shown in part (a), contains a block for each circuit group in the faulty unit. This group has an input but no output. The brackets enclose the entire unit. Take measurements at points calculated to eliminate nearly half of the remaining circuit groups with each measurement. A measurement at point C verifies an input to group 3. This eliminates groups 1 and 2 and moves the brackets as shown in part (b). Now measure at either D or E. A measurement at E shows no signal. This eliminates group 5 and moves the brackets as shown in part (c). Another measurement, at D this time, shows no signal. Group 3 has an input

## Troubleshooting Techniques

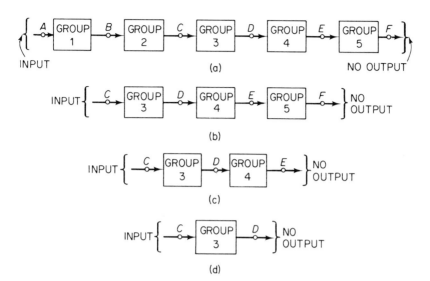

Figure 19-4. Bracketing Process

but no output. The brackets now enclose group 3 as the faulty circuit group.

This bracketing method of narrowing the problem area is very effective, and it is a time-saving technique. However, care must be exercised in constructing the chart. The example illustrated here is for linear signal paths with one input and one output. There are actually six types of signal paths, and the bracketing process must take these different paths into consideration, as illustrated in Figure 19-5.

Section (a) shows the *linear path* just described. The *divergent path*, section (b), has one input and two or more outputs. The *convergent path*, section (c), has two or more inputs and only one output. The *combination convergent–divergent path* has inputs that converge at a point and then spread out into multiple paths again. The *feedback path* is a variation of a divergent, a convergent, and a linear path. The *switching path* provides a choice between two or more linear paths. All paths must be considered during the isolation process. Otherwise, the troubleshooter is likely to be led to a false conclusion.

Once the faulty circuit group is determined, a few more readings will serve to isolate the faulty circuit in the group. In the case of the receiver in Figure 19-3, two groups contain only one circuit each. These are the RF section and the detector section. The frequency converter section has only two stages. The audio section is apparently the worst

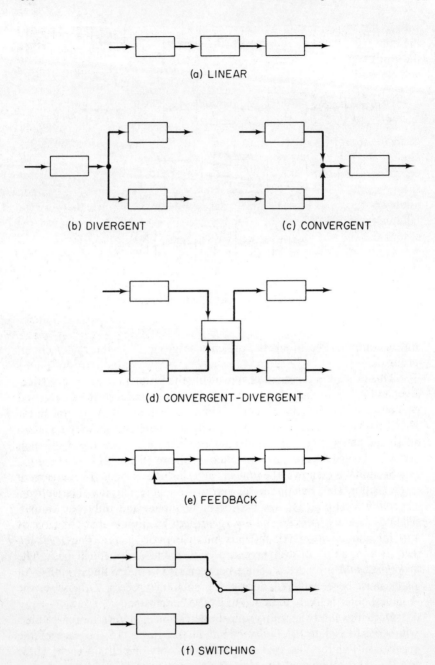

Figure 19-5. Types of Signal Paths

Troubleshooting Techniques

with its six stages. However, if the symptom is a complete loss of sound, the convergent and divergent paths make it very easy. It must be either the transformer or the speaker. A single reading at TP (test point) 6 will isolate the trouble to one or the other.

### 19.1.6 Step 6—Isolating the Faulty Component

Extensive use of logic and very little actual testing will generally carry the troubleshooter through the first five steps and isolate the faulty circuit. He is now ready for step 6—to determine exactly which of the components in that circuit has failed. Occasionally a quick visual inspection will accomplish this task. A loose connection, a bad solder joint, an open pin, a dark tube, or a cold transistor—all are possible failures that might be spotted by inspection. Such defects as charred resistors and shorted leads are also easy to spot.

More often the entire stage will appear perfectly normal, and additional testing will be required to ferret out the faulty component. Furthermore, even after one bad component is located, it cannot be assumed that it is the only one. An open resistor or a shorted capacitor may have been caused by excessive current. Therefore, the stage where the trouble exists must be subjected to extensive tests and measurements to make sure that all the trouble has been located.

Now is the time to bring out the schematic diagram of the faulty unit and analyze the faulty circuit. The schematic diagram provides a detailed picture of every component and its electronic relationship to the other components. When this information is reinforced with a knowledge of how this stage functions, it is a minor problem to perform detailed, comprehensive checks of the entire circuit. It is not a staggering task because any given stage has only a limited number of components. For example, if the trouble is in the audio-voltage-amplifier stage of a transceiver, the schematic diagram will be similar to that in Figure 19-6.

Since power is still on from the last measurements, the voltage readings are indicated as the most logical first measurement in step 6. The dc voltage values from all critical points to ground are clearly indicated on the schematic diagram. In this case those values are circled. The values of grid, cathode, plate, and applied voltage are quickly checked by either a voltmeter or an oscilloscope. Most troubleshooters prefer the oscilloscope at this point in the process. It has already been in use and is ready for action. A touch of the probe to tube pins 1 and 2 on each side of $R_{47}$ and the dc values are either verified or deviations are noted.

If one (or more) voltage(s) is wrong, why is it wrong? This question indicates that all resistances should be measured. For that matter,

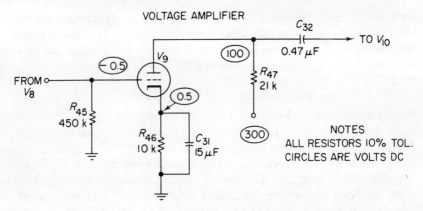

Figure 19-6. Sample Schematic Diagram of a Stage

even if all voltages are correct, the resistances should still be measured. After all, this is the faulty stage; something is definitely bad.

Before making a resistance measurement, the troubleshooter may want to verify his previous conclusion that the stage has an input signal but no output signal.

When all signal and voltage measurements have been completed, power is removed and resistances are measured with an ohmmeter. In the same-stage schematic diagram, there are no parallel resistors to interfere with the readings. Therefore, the measurement directly across each resistor should be the value indicated on the schematic ± the tolerance figure. In many cases, resistor measurements are more complicated. The schematic diagram then indicates vital resistance values from specified points to ground or between specified points. This enables a valid resistance measurement without the necessity of disconnecting all resistors from the circuit.

By this time the primary trouble has undoubtedly been found. But there are two more tests that should be made. While the power is off and the ohmmeter is handy, check the capacitors with the ohmmeter. Then check the tube or transistor. It is a small matter in most cases to remove the tube and check it with a tube checker. When the stage is transistorized, the transistor should be subjected to an in-circuit test.

The troubleshooter now knows which component has failed and whether there are secondary problems.

### 19.1.7 Repairing and Rechecking

Theoretically the troubleshooter's job is finished; he has found the trouble. And with a little practice, he will consistently locate troubles in less time than it takes to read the six steps of troubleshooting. But practically speaking, his job is not over because in most cases the troubleshooter is also the technician who does the repair. He now gets out his tools and removes all the faulty and marginal components and replaces them with identical (within-tolerance) components.

When the circuit is all back in one piece, each of the measurements must be taken again. Why? Because the repair work quite frequently causes other troubles to develop. After each part of the stage checks out to be right, the overall performance of the equipment must be checked with the repaired circuit in operation. If there is any deterioration in system performance, it is probably the result of the repaired section. If the performance has been degraded a noticeable amount, several adjustments may be required to bring it up to the specified standard.

## 19.2 Troubleshooting Charts

Nearly all systems contain some type of troubleshooting charts as part of the maintenance manuals. These charts are composed from theoretical troubles for the most part. In most cases the actual trouble will not appear on the charts. However, a study of these charts will be helpful in developing troubleshooting skills for new equipment. These charts list the primary method of fault detection, the symptoms, and several possible troubles that can cause the symptoms. A sample of such a chart is shown in Figure 19-7. This chart is for an amplifier stage in a transmitter that has built-in meters for monitoring the dc current for both control grid and plate. The symptoms are detected on one of these meters.

There are many aids to troubleshooting, but none are more valuable than knowledge of the equipment and the practice of a logical approach.

## 19.3 Malfunction Analysis

Engineers and technicians involved in equipment design seldom have an opportunity to work with complete systems. Rather, they spend a good deal of time designing, building, and testing individual stages. This is similar to the activity of a troubleshooter in isolating the faulty component in a circuit. This chapter is devoted to trouble analysis of individual stages. It is intended to improve the techniques of the experienced and to aid the beginner in developing sound troubleshooting skills.

A variety of sample circuits is used. These are common circuits, and they will be recognized by nearly everyone concerned. For this

| Meter | Symptoms | Possible Trouble |
|---|---|---|
| Grid current | Low reading | 1. Weak grid drive<br>2. Plate of previous stage mistuned<br>3. Bad tube |
| | Zero reading | 1. No grid drive<br>2. No RF in plate of previous stage<br>3. Bad tube |
| | High reading | 1. Open screen resistor<br>2. No screen voltage<br>3. No plate voltage |
| Plate current | Low reading | 1. Weak tube<br>2. Open grid resistor |
| | Zero reading | 1. Bad tube<br>2. No grid drive<br>3. Open screen resistor<br>4. No plate voltage<br>5. Open filament lead |
| | High reading | 1. No grid drive<br>2. Mistuned plate tank |

Figure 19-7. Troubleshooting Chart

reason the theory of operation and the normal functions are covered only briefly. The emphasis is on the symptoms and the possible troubles that can cause these symptoms.

Since each stage has a limited number of components, there are a limited number of malfunctions that are possible in any given stage. The techniques described in trouble analysis of these common circuits are applicable to all types of circuits, whether newly designed or established parts of a functional system.

### 19.3.1 Audio-Voltage Amplifier

Most radio and television receivers have an audio-voltage amplifier located between the detector and the power amplifiers. Its purpose is to accept the low-voltage signal from the detector and amplify it to provide sufficient drive for the power amplifiers. The active component in this circuit, and in most other circuits, may be either a bipolar transistor or a field-effect transistor. Figure 19-8 shows a circuit of each type to illustrate the similarities.

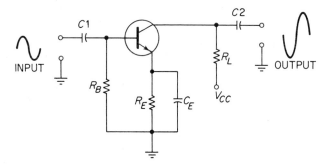

(a) BIPOLAR TRANSISTOR

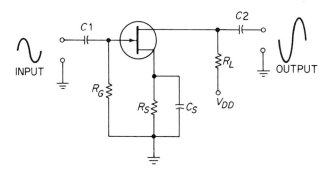

(b) FIELD EFFECT TRANSISTOR

Figure 19-8. Audio-Voltage Amplifiers

Notice that the two circuits are identical, with the exception of the amplifying device. In practice, of course, the voltage levels and component values will differ. The bipolar transistor is usually in a common-emitter configuration, although other arrangements are sometimes used. This is an *NPN,* but the *PNP* serves just as well. The field-effect transistor here is an *N*-channel in a common-source arrangement, but it is frequently seen as a *P*-channel.

Each of the circuits is a self-biasing, class A audio amplifier with degenerative feedback. Several different types of coupling and feedback arrangements may be expected. Circuit failure usually appears in one of three ways: no output, reduced output, or distorted output.

When the symptom is input with no output, a check of the dc voltages will usually pinpoint the defective component. If voltages are normal, the trouble is likely to be faulty coupling capacitors; frequently, this is a faulty solder connection. The fault may be an open-load resistor or an open-biasing resistor. In either case, the amplifier would be cut off. The amplifying device itself could also be bad, but this is less likely with this symptom.

The reduced-output symptom probably indicates a bad amplifying device. This is especially true if the output has gradually deteriorated over a period of time. A change in value of the biasing components sometimes causes this symptom. This alters the bias and affects the operational characteristics of the amplifier. Also, a reduced-load impedance will reduce the output by lowering the gain of the stage. An open bypass capacitor may be the problem, but this is the fault of a connection more often than the capacitor. An open lead or a cold solder joint to a capacitor produces a defective capacitor and results in excessive degeneration.

The distorted output is the usual symptom for too much gain. This can be caused by a number of faults. The amplifying device may have changed its characteristics, but other components are more likely. An open-biasing resistor will cause a loss of bias, with resultant excessive gain. Faulty decoupling capacitors may cause too much regenerative feedback. Open components in a degenerative feedback circuit may cause the same effect.

The recommended troubleshooting procedures are as follows:

1. Perform a close visual inspection.
2. Check the dc voltages.
3. Measure point-to-point resistances.
4. Check capacitors with ohmmeter.
5. Check the amplifying device.
6. Always check for secondary damage, and, after repair, perform an operational check.

### 19.3.2 Single-Stage Phase Splitter

The audio *phase splitter* is an amplifier that produces two outputs equal in amplitude and 180 degrees out of phase. A schematic diagram for a single-stage phase splitter is shown in Figure 19-9. This common emitter with one input and two outputs is only one of many versions of the phase splitter. This split-load arrangement, which produces equal and opposite outputs in respect to ground, is sometimes called a *phase inverter*. Failure in this circuit can cause one of four symptoms: (1) no output, (2) only one output, (3) reduced outputs, or (4) distorted outputs.

The no-output symptom can only be caused by a trouble that causes the amplifying device to cut off. In the example shown, any open resistor would disable the stage. The amplifying device may be bad, but one of the resistors is more probable.

Troubleshooting Techniques

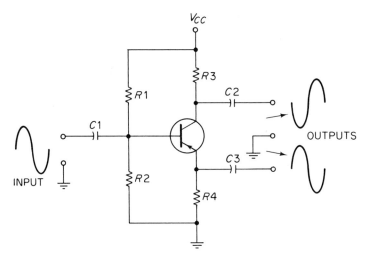

Figure 19-9. Single-Stage Phase Splitter

A loss of only one output leaves no room for doubt. The associated coupling capacitor is open. This is probably a faulty connection.

The reduced-output symptom is probably caused from an amplifying device that has weakened with age. A change of resistance in the input circuit could cause this symptom by changing the bias or reducing the amplitude of the input signal. Cold joints in the solder connections can cause attenuation that results in a low output.

The distorted output is caused from excessive gain. This may be due to a change in resistance in the biasing circuit. Peak clipping may result from a bias that is either too high or too low. A defective amplifying device can cause distortion, but the resistors are more probable.

### 19.3.3 Paraphase Amplifier

The *paraphase amplifier* is composed of two stages of amplification. Its purpose is to amplify an input signal and produce two outputs that are equal in amplitude and 180 degrees out of phase. It is sometimes called a *phase inverter,* and it is often used instead of a single-stage phase splitter. The outputs are used to drive power amplifiers. A transistorized paraphase amplifier schematic diagram is illustrated in Figure 19-10.

Each of the stages should be considered as a separate amplifier for trouble analysis. The first stage is a common amplifier with a certain amount of degeneration in the emitter circuit. It can produce the regular symptoms of no output, reduced output, or distorted output. The analysis and trouble isolation procedures are the same as described for voltage amplifiers.

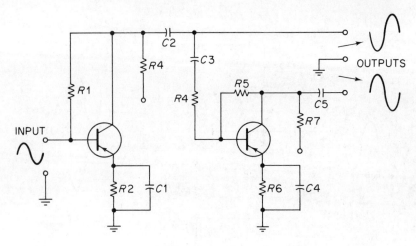

Figure 19-10. Paraphase Amplifier

The second stage is an amplifier, but the feedback is designed to cancel the gain. The input through $C_2$ should be approximately the same amplitude as, and 180 degrees out of phase with, the output. The symptoms are no output, reduced output, or distorted output. The analyses of the failure and troubleshooting procedures are the same as for voltage amplifiers.

### 19.3.4 Push–Pull Amplifier

The *push–pull amplifier* is normally the last stage in a receiver. It is the power amplifier that drives the speaker. This stage requires two inputs equal in amplitude and 180 degrees out of phase. The single output is a high-power, distortion-free signal. The schematic of a transistorized version is shown in Figure 19-11.

This is a class A amplifier consisting of two transistors connected back to back. It is a single-stage amplifier with two amplifying devices. The resistor section composed of $R_1$, $R_2$, and $R_3$ is a balance network to compensate for the fact that no two amplifying devices conduct exactly the same. The rheostat ($R_1$) can be adjusted for a balanced output.

The symptom of no output from a push–pull amplifier is very definitive. The amplifying devices can be eliminated at once. It is highly improbable that they would both go bad at the same time. This narrows the trouble to a component that will disable both the amplifying devices.

The reduced output normally signals failure of half the stage; that is, one amplifying device has no output. This is either a bad amplifying device or a bad component which disables one half of the push–pull stage. A biasing resistor could cause this problem.

# Troubleshooting Techniques

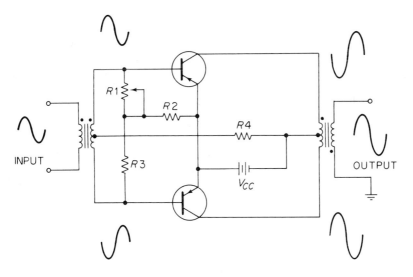

Figure 19-11. Push–Pull Amplifier

The distorted-output symptom is usually caused by improper bias resulting from components changing values.

### 19.3.5  Hartley Oscillator

The *Hartley oscillator* is a familiar circuit that comes in two versions: series and shunt. This analysis concerns a transistorized shunt-fed Hartley. A schematic for this oscillator is illustrated in Figure 19-12.

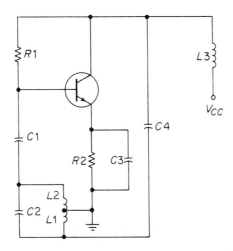

Figure 19-12.  Hartley Oscillator

The frequency-determining tank circuit is composed of $C_2$, $L_1$, and $L_2$. The feedback circuit consists of $C_4$ and $L_1$. $L_3$ is a radio-frequency choke that serves as the collector load. The oscillator is self-starting, and the feedback sustains the oscillations. Trouble in an oscillator circuit normally produces one of three symptoms: (1) no oscillation, (2) low-amplitude or unstable oscillation, or (3) oscillation at the wrong frequency.

Most of the possible troubles will cause oscillation to cease. Such conditions include an open or shorted transistor; an open or shorted capacitor: $C_1$, $C_2$, or $C_4$; or an open coil: $L_1$, $L_2$, $L_3$; or an open resistor: $R_2$. The transistor is likely to deteriorate with age, especially under high-temperature operating conditions. If $C_4$ opens, feedback ceases and stops oscillation. If $C_4$ shorts, the positive $V_{CC}$ to $L_1$ stops oscillation; this short-circuits the biasing circuit. If $C_1$ opens, there is no coupling from the tank to the transistor base, no amplification, no feedback, and no oscillation. If $C_1$ shorts, negative $V_{CC}$ through $L_2$ to the base causes the transistor to cut off and stops oscillation.

If $C_2$ opens or shorts, there is no tank circuit to oscillate, and this capacitor is not easy to check. There is no definite check except to disconnect the capacitor from the circuit. The tank is also destroyed by an opening in either $L_1$ or $L_2$. An opening in either $R_2$ or $L_3$ will disable the transistor and stop oscillation. In some cases opening $R_1$ will cause oscillation to cease, but this is improbable.

Instability comes in two forms: frequency and amplitude. Frequency instability is usually a symptom caused by poor connections, poor insulation, or changes in value of coils or capacitors. Amplitude or reduced output are symptoms that can be caused by improper bias, degenerating conditions, or an opened or shorted $C_3$.

Incorrect frequency is a common symptom, and this can be caused by many deteriorating conditions. Major frequency changes occur with changes in the transistor characteristics or changes in the value of the tank-circuit components.

Care must be exercised when checking oscillator circuits, and test instruments must have a common ground. Meters should be used with high-impedance probes, and the operational circuit must not be excessively loaded. Misuse of test instruments will cause faulty indications, and it may cause damage to the circuit.

### 19.3.6 Colpitts Oscillator

The *Colpitts oscillator* is another familiar circuit. The schematic diagram for one transistorized version is illustrated in Figure 19-13.

In appearance this Colpitts oscillator is very similar to the shunt-Hartley, but it is a more stable oscillator. The Colpitts is also capable of

Troubleshooting Techniques

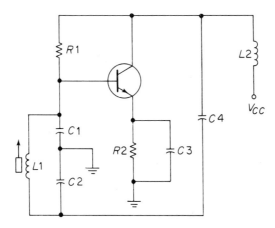

Figure 19-13. Colpitts Oscillator

reliable oscillation at frequencies higher than the Hartley is capable of. Since the Colpitts oscillator is capable of producing a sine-wave output of constant amplitude and frequency from audio through the high-RF range, it is a very popular oscillator for receiver local oscillators, signal sources in signal generators, and for many applications requiring variable-frequency oscillators.

The frequency is determined by a tank circuit composed of a tunable coil and two capacitors. The feedback to sustain oscillation is supplied through a capacitor voltage divider. The voltage across $C_1$ is the input to the transistor, and the quantity of feedback is determined by the ratio between $C_1$ and $C_2$.

The symptom, no oscillation, will appear with any failure that disables either the transistor or the tank circuit. An open or a short of the transistor or in $C_1$, $C_2$, or $C_4$ will stop oscillation. A bad transistor stops all current. Opening $C_4$ stops feedback to the tank circuit. Shorting $C_4$ applies dc to both sides of the tank circuit. Opening or shorting $C_1$ or $C_2$ disables the tank circuit. An open in either coil or in $R_2$ will stop oscillation. Opening $L_1$ disables the tank circuit. Opening $L_2$ or $R_2$ disables the transistor circuit.

Instability in amplitude or reduced output is likely to occur as a result of improper biasing. This is usually caused by characteristic changes due to extreme temperatures. Reduced amplitude means too much bias or excessive degeneration, and this can be caused by increased resistance in the bias circuit ($R_1$ or $R_2$) or an open in $C_3$. A leaky coupling capacitor, $C_4$, will reflect a resistance through the inductance of the tank that appears between the base and collector of the transistor; this also changes bias.

Frequency instability indicates problems with the tank circuit. This can be caused by poor mechanical connections or deteriorated insulation. Partial shorts between windings of $L_1$ can also cause frequency instability. Improper conduction of a faulty transistor may cause frequency shifts, but this will more than likely appear as amplitude instability.

Incorrect output frequency is not as common as with the Hartley oscillator. When it does happen, it is a result of deteriorating conditions. The deterioration may be in the tank-circuit components, transistor, or coupling. A change in parameters causes a more pronounced frequency shift when operating at higher frequencies.

### 19.3.7 Clapp Oscillator

The *Clapp oscillator* is considered to be a variation of the Colpitts. It is used to produce a sine wave within the RF range. This oscillator is frequently employed in signal generators and as variable-frequency oscillator for general use over the high and very high frequency ranges. Figure 19-14 is a schematic diagram of a transistorized Clapp oscillator.

The frequency determining device is a series resonant circuit composed of $L_1$ and $C_1$. $C_2$ and $C_3$ form a voltage divider that determines the amplitude of the feedback voltage. The transistor has fixed bias de-

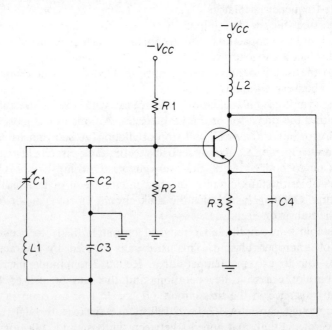

Figure 19-14. Clapp Oscillator

veloped by $R_1$ and $R_2$. $R_3$ and $C_4$ are emitter components to prevent instability and degeneration. The tuned circuit is designed for a high $Q$ under load, which gives the Clapp oscillator greater stability than the basic Colpitts.

The no-oscillation symptom is an indication of an opened or shorted condition that disables either the transistor or the tuned circuit. Any opened component, with the exception of $C_4$, can cause this symptom; this includes the transistor. Shorting of capacitor $C_1$, $C_2$, or $C_3$ will disable the tuned circuit and stop oscillation. Poor connections may effectively produce one or more of these conditions and result in no oscillation.

Reduced or unstable amplitude of oscillation is generally caused by a change in bias resistance or a defective bypass capacitor in the emitter circuit. Loss of gain due to transistor deterioration may also cause a reduced or unstable output.

Frequency instability is generally caused by poor connections that affect the tank-circuit components. Incorrect frequency can be caused by changes in parameters that vary the distributive capacitance. Variation in the inductance of the tuned circuit will cause a frequency shift. Shorting of a few coils on the inductor will change the inductance. A change in the value of the tuning capacitor would cause frequency shifts also, but this probably would not be noticed until the capacitor failed completely.

### 19.3.8 Crystal Oscillator

There are several types of *crystal oscillators*. The *Colpitts,* the *Miller,* and the *Pierce* are among the most popular. Crystal oscillators are likely to be used in any circuit that requires a radio-frequency signal of extreme stability. The example used here is a Colpitts crystal oscillator. It is generally employed at the higher radio frequencies. Figure 19-15 is a schematic of a transistorized version of this oscillator.

The Colpitts crystal oscillator uses external capacitive divider feedback to sustain oscillation of the crystal. $C_1$ and $C_2$ provide this feedback. Another voltage divider composed of $R_1$ and $R_2$ supplies a fixed bias for easy starting. The quartz crystal, operating on the piezoelectric effect, produces oscillation at a stable frequency. The frequency of the oscillator is not limited to the fundamental crystal frequency; it may be equivalent to any one of several harmonics of this basic frequency.

A symptom of no oscillation is an indication of a trouble that stops the transistor from conducting, stops feedback, or disables the crystal itself. The easiest way to check a suspected crystal is to substitute one that is known to be good. This substitution can be made without breaking the circuit.

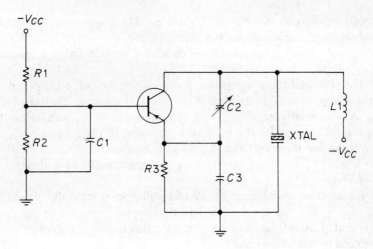

Figure 19-15. Colpitts Crystal Oscillator

An open-bias resistor will prevent starting by keeping the transistor cut off. The same is true of the collector load inductor, $L_1$. If either $C_2$ or $C_3$ opens, the feedback network is inoperative, and this stops oscillation.

Reduced amplitude or instability of either amplitude or frequency can be caused by deterioration of parts due to age or extremes of temperature. A frequent cause of such a symptom is a damaged crystal. Changes in the characteristics of the crystal may be caused by age, partial fracture, dirty crystal holder, or improper control of temperature and moisture. Any deteriorating condition that reduces the conduction of the transistor can also cause some form of reduced or unstable output.

Oscillation at an incorrect frequency is more than likely due to a faulty crystal. Changes in the value of the feedback capacitors, $C_2$ and $C_3$, could cause this problem, but, since they are tunable, the result would probably go unnoticed until one of them failed completely. Simple measurements, such as those for voltage, resistance, and continuity, will generally isolate the faulty component.

# Bibliography

Bell, David A., *Fundamentals of Electronic Devices*. Reston, Va.: Reston Publishing Company, 1975.

Berens, Jack, and Berens, Stephen, *Understanding and Troubleshooting Solid State Electronic Equipment*. Radnor, Pa.: Chilton Book Company, 1969.

Coombs, Clyde F., Jr., *Printed Circuits Handbook*. New York: McGraw-Hill Book Company, 1967.

Corning, John C., *Transistor Circuit Analysis and Design*. Englewood Cliffs, N.J.: Prentice-Hall, Inc., 1965.

DeWaard, Henrick, and Lazarus, David, *Modern Electronics*. Reading, Mass.: Addison-Wesley Publishing Company, Inc., 1966.

Gaddis, Ben W., *Troubleshooting Solid State Power Supplies*. Blue Ridge Summit, Pa.: Tab Books, 1972.

General Electric, *SCR Manual*. Syracuse, N.Y.: General Electric Company, 1972.

General Electric, *Transistor Manual*. Syracuse, N.Y.: General Electric Company, 1969.

Ghausi, Mohammed S., *Electronic Circuits*. New York: Van Nostrand Reinhold Company, 1971.

Grabinski, Joseph, *Electronic Power Supplies*. New York: Holt, Rinehart and Winston, Inc., 1969.

Graf, Rudolf F., *Electronic Design Data Book*. New York: Van Nostrand Reinhold Company, 1971.

Hemingway, T. K., *Electronic Designers Handbook*. London: Business Books Ltd., 1966.

Hildreth, Hugh, *Electrical Engineering Circuits*. New York: John Wiley & Sons, Inc., 1965.

Horowitz, Mannie, *How To Build Solid State Audio Circuits*. Blue Ridge Summit, Pa.: Tab Books, 1972.

Hughes, L. E. C., *Handbook of Electronic Engineering*. Cleveland, Ohio: CRC Press, 1967.

Jones, Edwin R., *Solid State Electronics*. Scranton, Pa.: Intext Educational Publishers, 1971.

Lenk, John D., *Handbook of Practical Electronic Tests and Measurements*. Englewood Cliffs, N.J.: Prentice-Hall, Inc., 1969.

Lenk, John D., *Handbook of Simplified Solid State Circuit Design*. Englewood Cliffs, N.J.: Prentice-Hall, Inc., 1971.

Malvino, Albert Paul, *Transistor Circuit Approximations*. New York: McGraw-Hill Book Company, 1969.

Olson, Harry F., *Modern Sound Reproduction*. New York: Van Nostrand Reinhold Company, 1972.

Pullen, Keats A., *Handbook of Transistor Circuit Design*. Englewood Cliffs, N.J.: Prentice-Hall, Inc., 1961.

RCA, *Transistor, Thyristor and Diode Manual*. Harrison, N.J.: RCA Electronic Components, 1969.

Robinson, Vester, *Handbook of Electronic Instrumentation, Testing, and Troubleshooting*. Reston, Va.: Reston Publishing Company, 1974.

Robinson, Vester, *Solid State Circuit Analysis*. Reston, Va.: Reston Publishing Company, 1975.

Shea, Richard F., *Amplifier Handbook*. New York: McGraw-Hill Book Company, 1966.

Towers, T. D., *Transistor Television Receivers*. New York: John F. Rider, 1963.

Wheeler, Gershon J., and Tripp, Arley L., *Essentials of Electronics*. Englewood Cliffs, N.J.: Prentice-Hall, Inc., 1971.

# Index

AC
  alpha, 29
  beta, 29
Acceptor material, 1
Adjustable output regulator, 352
Alpha, 28, 41
Alternating current gain, 40
Ambient temperature, 97
Amplifier modulator, 265
Amplifiers, 125, 243
  audio frequency, 125
    class A, 180
    class AB, 189
    class B, 186
  bipolar, 127
  JFET, 155
  MOSFET, 165
  multistage, 208, 220
  radio frequency, 243
Amplifying device, 287
Amplitude modulator, 265
Applications of:
  parameters, 109
  RC filters, 211
  transformer coupling, 219
Armstrong RF oscillator, 303, 305
Astable multivibrator, 315
Audio frequency circuits, 125
  amplifiers, 125
    bipolar, 127
    JFET, 155
    noise, 196
    power, 176
  oscillators, 125
    keyed, 289
    lag line, 293
    twin-T, 291

Audio frequency circuits (*Cont.*)
  testing, 369
  malfunction analysis, 394
Avalanche current, 8
Avoiding false indications, 364

Bandpass, 245
  filter, 211
Band reject, 245
Bandwidth, 240, 248, 249, 270, 365
Base injection, 275
Basic
  FET circuits, 65
  JFET amplifier, 155
  MOSFET amplifier, 165
  parameters, 99
  series regulator, 347
Bass control, 237
Bench testing, 359
Beta, 28, 41
  AC, 29
Biasing
  bipolar, 26, 36, 129
  diodes, 4
  JFET, 50, 55, 72
  MOSFET, 74
  with transformers, 189
Bipolar Colpitts RF oscillator, 295
Bipolar transistor, 24
  amplifier, 127
  biasing, 26, 36, 129
  circuit configurations, 29
  current
    carriers, 26
    distribution, 27
    electron, 25
    leakage, 29, 38, 117

Bipolar transistor (*Cont.*)
  ratios, 28
  noise, 204
  NPN, 24
  parameters, 42, 99, 103
  PNP, 25
  switching characteristics, 367
  symbols, 45
  tests, 38
Bistable multivibrator, 319
Blocking oscillator, 312
Blue noise, 200
Breakdown voltage, 16
Bridge rectifier, 333

Capacitance measurement, 362
Capacitor input filter, 334
Channel
  depletion MOSFET, 61
  depletion mode, 62
  enhancement MOSFET, 58
  enhancement mode, 662
Charge time, 307
Choosing a transistor, 99
Clapp oscillator, 402
Class of operation
  A, 180
  AB, 189
  B, 186
Collector
  base junction, 25
  current, 84, 95
  injection, 266
Colpitts oscillator, 295, 298, 300, 400
Common
  base, 30
    equivalent circuit, 109
    load line, 140
    logical equations, 139
  collector, 34
    equivalent circuit, 113
    logical equations, 138
  drain, 70
  emitter, 31, 145
    equivalent circuit, 111
    load line, 146
    logical equations, 136
  gate, 65
  source, 68
Complementary circuit, 188, 190, 225
Contour curves, 204
Control circuits
  bass, 237
  gain, 233
  treble, 234
Converters, frequency, 276
Corner frequency, 209, 210, 211, 216

Coupling, 208
Crossover distortion, 187
Crystal controlled oscillators, 403
  Armstrong, 305
  Colpitts, 298
Current
  capacity, 330
  carriers, 26
  feedback, 230
  gain, 117
    vs frequency, 116
  latching, 20
  Zener, 9
Cut-in voltage, 5
Cutoff characteristics, 97
Curves
  contour, 204
  power, 177, 183

Data
  groups, 83
  sheets, 79, 81, 86, 88
Darlington
  compound, 223
  dual, 226
Decade boxes, 124, 130
Degenerative feedback, 227, 229
Degree of modulation, 268
Demodulators, 278
Dependent circuits, 233
Depletion
  mode, 62
  zone, 10
Deriving logical equations, 135
Design examples
  amplifiers
    AF voltage, 149
    basic MOSFET, 172
    class AB, 190
    MOSFET, 172
  regulators
    basic series, 348
    electronic shunt, 347
    Zener diode, 344
Detector
  diode, 278, 279
  ratio, 283
  transistor, 280
Device capabilities, 87
Differential amplifier, 224
Diodes
  PN junction, 3
  tunnel, 10
  steering, 322
  symbols, 21
  varactor, 14

Index

Diodes (*Cont.*)
    Zener, 3, 8
        regulator, 343
        sawtooth generator, 311
Direct
    coupling, 208, 220
    current
        characteristics, 85, 96
        gain, 40
Direction of current, 3
Discharge time, 307
Discriminator, 281
Distortion in amplifiers
    bipolar audio, 152, 154
    class A, 184
    class AB, 195
    class B, 187
    feedback circuit, 240
    MOSFET, 175
Distribution of current, 27
Distributive capacitance measurement, 363
Dividers, voltage, 338
Donor material, 1
Doubler, voltage, 336
Drain
    resistance, 57
    to source voltage, 207
Dual output regulator, 355

Effect of frequency on parameters, 115
Efficiency of
    frequency multipliers, 272
    power transformers, 327
    transformer coupling, 214
Electrical analogy of thermal circuits, 118
Electron current in transistors, 25
Electronic
    series regulator, 350
    shunt regulator, 345
    switches, 3
Emitter
    base junction, 25
    injection, 275
Enhancement mode, 62
Enhancing symptoms, 382
Equations
    feedback, 228
    logical, 135
Equivalent
    circuit
        bipolar amplifier, 127
        diode, 7
        filters, 211
        JFET amplifier, 158, 160
        MOSFET amplifier, 168
    noise generator, 203

Feedback, 227, 272, 288
    in IF amplifiers, 258
    in oscillators, 288
    regulator, 352
FET, 49
    noise, 205
    switching characteristics, 367
    transfer characteristics, 53, 55
    symbols, 77
Field effect transistor, 49
Filter capacitor, 334
Filters, RC
    applications, 213
    bandpass, 211
    high pass, 209
    low pass, 210
    power supply, 334
Fixed component spare parts, 124
Four terminal parameters, 42, 99
Frequency
    corner, 209, 210, 211, 216
    determining device, 288
    doubler, 271
    modulators, 268
    multipliers, 271
    resonant, 245
    response, 153, 218, 369
    stability, 289
Full-wave rectifier, 332

Gain
    alpha, 28, 41
    antenna, 378
    bandwidth product, 116
    beta, 28, 41
    bipolar amplifier, 152
    control, 233
    common emitter, 146
    current, 117, 134
    direct current, 40
    JFET amplifier, 161, 164
    MOSFET amplifier, 175
    power, 134
    voltage, 134
    with feedback, 241
Gate, common, 65
General solid state symbols, 44
Generator
    noise, 203
    sawtooth, 309, 311
    signal, 122, 123, 152, 285

Half-wave rectifier, 330
Harmonic distortion, 184, 370
Hartley oscillator, 399
Heat sinks, 118

Heavy current voltage regulator, 356
Heterodyning, 272, 273
High
  frequency
    characteristics, 85, 97
    problems, 261
  level modulation, 265
  pass filter, 209, 216
Holding current, 16
Hybrid
  equivalent analysis, 132
  parameters, 42, 99, 103

$I_{CBO}$ test, 39
$I_{CEO}$ test, 39
Impedance
  antenna, 377
  in feedback circuits, 240
  matching, 128
Inductance measuring, 362
Injection
  base, 274
  emitter, 275
Input
  capacitor filter, 334
  impedance
    audio circuits, 372
    bipolar amplifiers, 132
    JFET amplifiers, 160
    MOSFET amplifiers, 169
  output regulator, 354
  resistance, 58
Isolating faulty
  components, 391
  function, 385
Instability, 220, 222
Insulated gate FET
  JFET, 49
  MOSFET, 58
Intermediate frequency
  amplifiers, 257
  transformers, 258
Intermodulation distortion, 185
Internal resistance, 136
Interpreting data sheets, 95

Junction
  capacitance, 15
  collector-base, 25
  emitter-base, 25
  FET, 49
    basic amplifier, 155
    load line, 162
    measuring, 368
    video amplifier, 263
  NPN, 25
  PN, 3

PNP, 24
Lag line oscillator, 293
Latching current, 20
Leakage current
  bipolar, 29, 38
  collector, 38
  $I_{CBO}$, 39
  $I_{CEO}$, 39
  MOSFET, 58
Light sensitive SCR, 17
Limitation of power, 176
Listing probable faulty function, 385
Load line, 140
  audio power amplifier, 179
  class A amplifier, 183
  common
    base, 141
    emitter, 146
  JFET, 162
  MOSFET, 170
Load, transformer, 182
Loading of resonant circuit, 249
Local oscillator, 276
Localizing trouble, 386
Logical equations, 135
Low
  frequency problem, 259
  level modulation, 265
  output, 195
  pass filter, 210, 216

Malfunction analysis, 393
Matching impedance, 128
Maximum ratings, 83, 87
Measuring
  antennas, 376
  audio circuits, 369
  bipolar switching characteristics, 367
  circuit characteristics, 361
  FET switching characteristics, 368
  magnetic components, 373
  noise ratios, 197
Mixers, 272
Mode
  depletion, 62
  enhancement, 62
Model circuits
  basic JFET, 155
  bipolar amplifier, 127
  class AB amplifier, 191
  common emitter, 145
  MOSFET, 165, 172
Modulation, 265
MOSFET, 58
  basic circuits, 65
  basic amplifier, 165
  biasing, 74

# Index

MOSFET *(Cont.)*
  RF oscillator, 300
Multimeter, 122, 123
Multiple load voltage divider, 340
Multipliers, frequency, 271
Multistage amplifiers, 208
Multivibrators
  astable, 315
  bistable, 319
  monostable, 317

Negative
  feedback, 227
  resistance, 10
Noise
  audio amplifier, 196
  bipolar transistor, 204
  FET, 205
  generator, 203
  ratio, 196
  spectrum, 199
NPN transistor, 25

Open circuit parameters, 99
Operating
  frequency, maximum, 116
  point
    common base, 142
    common emitter, 146, 150
    MOSFET, 170, 173
    tunnel diode, 11
Oscillator modulator, 267
Oscillators
  audio, 125
  local, 276
  principles, 287
Oscilloscope, 122, 123, 152
Output
  admittance, 57
  feedback regulator, 352
  impedance
    bipolar, 133
    measuring, 373
  regulator, dual, 355

Parallel resonant circuits, 245
  measuring, 361
Parameters
  application, 109
  conversion, 105
  effect of
    frequency, 115
    temperature, 117
  four terminal, 42, 99
  hybrid, 42, 103
  open circuit, 99
  short circuit, 99, 102

Parameters *(Cont.)*
  transistor, 79
Paraphase amplifier, 397
Partition noise, 199, 200
Peak inverse voltage, 330
Phase relation in transformers, 329, 375
Pink noise, 200
PN junction, 3
PNP transistor, 24
Power
  across load, 182
  amplifiers
    audio, 125, 176
    RF, 244
  capabilities of transformer coupling, 220
  conversion circuits, 327
  curve, 177, 183
  dissipation, 117
  frequency multiplier, 272
  gain, 134
  limitations, 120, 176
  ratings, 84, 177
  rectifiers, 329
  supply
    filters, 334
    regulated, 122, 123
  transformer, 327
Practical biasing, 36
Principles of
  feedback, 227
  oscillation, 287
Punch through voltage, 41
Push-pull
  amplifier, 398
  distortion, 187

Radio frequency amplifiers, 243
  IF, 244, 257
  preamplifiers, 244, 254
  power, 244, 250
  video, 243, 244, 259
Radio frequency oscillators, 295
  Armstrong, 303
    crystal controlled, 305
  bipolar Colpitts, 295
    crystal controlled, 298
  MOSFET Colpitts, 295
Ratings, maximum, 83, 87
Ratios
  bipolar current, 28
  detector, 283
  signal to noise, 196
RC
  coupling, 208
  transients, 307

RC (*Cont.*)
    time constants, 308
Reactance oscillator modulator, 269
Rechecking, 392
Recognizing symptoms, 381
Rectifiers
    power, 329
    silicon, 15
        controlled, 15
    voltage, 343
Regenerative feedback, 227, 229
Region, Zener, 8
Regulated power supply, 122, 123
Regulators, voltage, 343
Relaxation oscillators, 285, 307
Repairing, 392
Resistance
    diode, 6
    drain, 57
    input, 57, 58
    internal, 136
    negative, 10
    thermal, 118
Resonant circuits
    parallel, 245, 361
    series, 245, 363
Results of feedback, 240
Reverse blocking current, 16
Rolloff frequency, 209, 210
Runaway, thermal, 16, 29

Sample circuits
    amplifiers
        bipolar audio, 127
        bipolar video, 262
        class A, 183
        class AB, 190
        class B, 186
        common emitter, 145, 149
        JFET, 155, 263
        MOSFET, 165, 172
        power, 183, 251
        preamplifier, 255
    frequency
        converter, 276
        mixer, 274, 275
        multiplier, 271
    oscillators
        Armstrong RF, 303, 305
        astable multivibrator, 317
        bistable multivibrator, 319
        blocking, 312
        Colpitts RF, 295, 298
        keyed audio, 289
        lag line, 293
        monostable multivibrator, 317
        MOSFET Colpitts RF, 300

Sample circuits (*Cont.*)
    sawtooth generator, 309, 311
    twin-T audio, 291
    rectifiers
        bridge, 333
        full-wave, 332
        half-wave, 331
    regulators
        basic series, 348
        dual output, 355
        electronic series, 350
        electronic shunt, 346
        for heavy current, 357
        Zener diode, 343
Second stage bias, 221
Semiconductors, 1
Selectivity, 248
Sensitivity, 240
Series
    diode detector, 278
    output amplifier, 224
    regulators
        basic, 347
        electronic, 350
    resonant circuits, 245, 363
Short circuit parameters, 99, 102
Shot noise, 199
Shunt
    diode detector, 279
    regulators
        electronic, 345
        Zener, 343
Sidebands, 270
Signal
    generation circuits, 285
    generator, 122, 123, 152
    swing in class A, 180
    to noise ratio, 196
Silicon
    controlled rectifier, 15
    rectifier, 15
Single
    load voltage divider, 338
    stage phase splitter, 396
Sinusoidal oscillators, 285, 287
Six step troubleshooting procedure, 380
Small signal characteristics, 84
Spare parts, 124
Speaker measurement, 374
Specifications, 149
Static
    characteristics
        common base, 31, 32
        common emitter, 33
    values in troubleshooting, 152, 174
Stability, 229, 289
Switches, electronic, 3

# Index

Switching characteristics, 85, 97, 367
Symbols
  bipolar transistor, 45
  diode, 21
  FET, 77
  solid state, general, 44
  thyristor, 21
  triac, 23
  tunnel diode, 22

Tank circuit, 245
Temperature, 84, 96, 97
Test
  circuits, 152, 172, 174
  equipment, 122, 359
Testing and troubleshooting, 152, 153, 154, 174, 175, 194, 263, 359, 380
Thermal
  circuits, 79, 118
  noise, 199
  resistance, 118
  runaway, 16, 29
Thyristors, 3, 17
  symbols, 21
Time constants, 308
Transients, RC, 307
Transformer
  coupling, 208, 214
  phase relation, 375
  power, 327
  regulation, 376
Transconductance, 56
Transfer characteristics, JFET, 53, 55
Transistor
  alpha, 28, 41
  bipolar biasing, 26, 36
  beta, 28, 41
  current, 25
  detector, 280
  IGFET, 49
  JFET, 49

Transistor (*Cont.*)
  MOSFET, 58
  parameters, 79
  tester, 122
  tests, 38
Treble control, 234
Triac, 17
  symbols, 23
Tripler, voltage, 337
Troubleshooting charts, 393
Tunnel
  diode, 10
  rectifier, 12
  symbols, 22
Tunneling, 10
Twin-T oscillator, 291

Unijunction sawtooth generator, 309

Varactor diode, 14
Variable components, 124
Video amplifiers, 243, 244, 259
Voltage
  amplifiers, 125, 127
  cut-in, 5
  dividers, 338
  drain to source, 207
  feedback, 230
  gain, 134, 161
  maximum, 84, 87
  multipliers, 335
  peak inverse, 330
  punch through, 41
  regulators, 343
  Zener, 9

Wide band amplifiers, 259
White noise, 199

Zener diode, 8, 9
  regulator, 343